Cell Biology and Bioengineering: From Concepts to Applications

Cell Biology and Bioengineering: From Concepts to Applications

Edited by **Samantha Granger**

SYRAWOOD
PUBLISHING HOUSE

New York

Published by Syrawood Publishing House,
750 Third Avenue, 9th Floor,
New York, NY 10017, USA
www.syrawoodpublishinghouse.com

Cell Biology and Bioengineering: From Concepts to Applications
Edited by Samantha Granger

International Standard Book Number: 978-1-68286-006-9 (Hardback)

Contents

Permissions

List of Contributors

Preface

The main aim of this book is to educate learners and enhance their research focus by presenting diverse topics covering this vast field. This is an advanced book which compiles significant studies by distinguished experts. This book addresses successive solutions to the challenges arising in the area of application, along with it; the book provides scope for future developments.

This book attempts to provide an in-depth insight into the emerging concepts and applications of bioengineering and cell biology. The topics compiled herein, such as gene therapy, cell and tissue culture, molecular biology, embryo-genetics, bioinformatics, etc. are bound to provide a thorough understanding of these two prominent fields. The researches and case studies included in the book give a broad overview of the current advancements in these fields. It will help new researchers by foregrounding their knowledge in these subjects.

It was a great honour to edit this book, though there were challenges, as it involved a lot of communication and networking between me and the editorial team. However, the end result was this all-inclusive book covering diverse themes in the field.

Finally, it is important to acknowledge the efforts of the contributors for their excellent chapters, through which a wide variety of issues have been addressed. I would also like to thank my colleagues for their valuable feedback during the making of this book.

Editor

Microsatellite DNA Marker Analysis Revealed Low Levels of Genetic Variability in the Wild and Captive Populations of *Cirrhinus cirrhosus* (Hamilton) (Cyprinidae: Cypriniformes)

Md. Abul Hasanat[1], Md. Fazlul Awal Mollah[1] and Md. Samsul Alam[1*]

[1]*Department of Fisheries Biology and Genetics, Bangladesh Agricultural University (BAU), Mymensingh 2202, Bangladesh.*

Authors' contribution

This work was carried out in collaboration between all authors. Authors MAH, MFAM and MSA designed the study. Authors MAH and MSA performed the statistical analysis, author MAH performed the literature searches, wrote the protocol and the first draft of the manuscript. Authors MFAM and MSA managed the analyses of the study. Finally, all authors read and approved the final manuscript.

Editor(s):
(1) Cosmas Nathanailides, Department of Aquaculture & Fisheries, TEI of Epirus, Greece.
(2) Kuo-Kau Lee, Department of Aquaculture, National Taiwan Ocean University, Taiwan.
Reviewers:
(1) Anonymous, Italy.
(2) Anonymous, India.
(3) Anonymous, Brazil.

ABSTRACT

Aims: To reveal the genetic variability of wild and captive populations of the Indian major carp, mrigal (*Cirrhinus cirrhosus*) based on microsatellite DNA markers analysis.
Study Design: Three rivers namely the Halda, the Padma and the Jamuna were selected under wild population category and three hatcheries such as Brahmaputra Hatchery of Mymensingh, Raipur Government Hatchery, Luxmipur, and Sonali Hatchery, Jessore were selected under the captive population category.
Place and Duration of the Study: Department of Fisheries Biology and Genetics, Bangladesh Agricultural University, Mymensingh from July 2005 to June 2008.
Methodology: DNA was extracted from fin clips of a total of 180 fish, 30 from each of the six populations. Five microsatellite markers MFW1, MFW2, MFW17, Barb54 and Bgon22 were

Corresponding author: Email: samsul.alam@bau.edu.bd

amplified by polymerase chain reaction for each DNA sample and resolved on denatured polyacrylamide gel and visualized by silver nitrate staining.

Results: Three of the five loci were found to be polymorphic in all the six populations. The observed (H_O) and expected heterozygosity (H_E) ranged from 0.233 to 0.633 and 0.406 to 0.664 respectively. The F_{IS} values ranged from 0.032 to 0.635 indicating deficiency in heterozygosity. Except the Raipur Hatchery stock, the other five populations showed nonconformity to Hardy-Weinberg Expectation at least in one locus. Significant population differentiation was observed between the Halda-Jamuna, Jamuna-Brahmaputra Hatchery, Jamuna-Raipur Hatchery and the Padma-Raipur Hatchery population pairs. The UPGMA dendrogram based on genetic distances resulted in two major clusters: the Halda river and the Raipur Hatchery population were in one cluster and the remaining four populations were in the other cluster.

Conclusion: The study, as a whole, revealed low levels of genetic variation in terms of allelic richness and heterozygosity in the three major rivers and three selected hatchery stocks of *C. cirrhosus* in Bangladesh.

Keywords: Genetic variation; microsatellite; major carp; Cirrhinus cirrhosus.

1. INTRODUCTION

Bangladesh with its neighboring countries in South Asia is the home to rich genetic resources of Indian major carps which includes catla (*Catla catla*), rohu (*Labeo rohita*), mrigal (*Cirrhinus cirrhosus*) and kalibaus (*Labeo calbasu*). Due to their complementary biological and ecological characteristics in efficiently using natural foods produced in different strata of a water body, these species have insured their positions as integral components of carp polyculture systems long been practiced in Bangladesh. Mrigal is considered as the third popular major carp species next to rohu and catla. The world aquaculture production of *C. cirrhosus* was 378622 MT in 2010, which ranked 24th among all aquaculture species [1]. Additional production is also obtained from natural sources through capture fishery.

Fish production largely depends on the availability of good quality fish seed irrespective of farming systems. The major carp populations in the natural water bodies have declined alarmingly due to habitat destruction and overfishing over the last few decades [2]. On the other hand, a high demand of carp seed for aquaculture and availability of low cost technology for induced breeding led to establishment of as many as 900 carp hatcheries across the country [3]. Therefore, dependence on nature for Indian major carps' seed has been totally replaced by hatchery produced seeds. The hatchling production in the hatcheries was increased from 4962 Kg in 1985 to 629175 kg in 2011 while the quantity of spawn collected from natural sources was reduced from 19,362 kg in 1980 to 4370 kg in 2011 [3]. It is a good sign as a whole for aquaculture industry but the deterioration of quality at the same time remains as a great concern for sustainable development as little attention has been paid to maintain genetic quality. In brief, two scenarios regarding the status of Indian major carps are evident: (1) sustained reduction in natural populations due to habitat destruction through continued siltation and over fishing, (2) intensification in fry production in the hatcheries giving little attention to the genetic quality of the stocks. Either of these two phenomena have profound impact on the genetic constitution and quality of the populations. Molecular marker analysis of samples from carp hatcheries revealed extensive hybridization among the major carp species in one hand and loss of genetic variation due to inbreeding and genetic drift on the other [4-7] and the lower growth performances of the hatchery produced seeds are frequently discussed in the sector. The genetic variation in a population is an important resource that needs to be preserved for maintaining fitness of a population to adapt to changing environments [8]. Therefore, documentation of intra-specific genetic diversity in the existing populations is essential to develop effective plans for future breeding program and conservation of the gene pools. Some attempts were made in Bangladesh to know the genetic status of different river and hatchery stocks of catla (*Catla catla*) [5,9] and rohu (*Labeo rohita*) [4,7,10,11]. However, despite of having similar importance as an integral component of carp polyculture systems, the genetic structure of mrigal (*C. cirrhosus*) in Bangladesh is less known Hasanat et al. [12] using allozyme markers reported lower genetic variability in the hatchery stocks of mrigal (*C. cirrhosis*) compared to the river counterparts.

Microsatellites, also called simple sequence repeats (SSR) markers, due to their ubiquitous presence in the genome and hypervariability in nature, have already been established as powerful tools for genetic characterization of plants and animals and have practical implications in the field of fisheries and aquaculture [13,14,6]. Microsatellites are found to show considerable evolutionary conservation allowing cross-species amplification as an alternative strategy for genetic characterization of a species using primers developed from other closely related species [15,16]. Based on characterization of polymorphic heterologous microsatellite markers we report here low levels of genetic variability within three wild and three representative captive populations of *C. cirrhosis* in Bangladesh.

2. MATERIALS AND METHODS

2.1 Sample Collection and Extraction of Genomic DNA

Three main river sources such as the Halda (HR), the Jamuna (JR) and the Padma (PR) were chosen as sources of wild *C. cirrhosus*. The spawn samples were collected from the rivers and reared in separate earthen nursery ponds of Bangladesh Agricultural University, Mymensingh campus until they attained sizes of ~7 cm. On the other hand, for samples of hatchery origin, three hatcheries were selected from three regions. Those were Brahmaputra Hatchery (BH) of Mymensingh (central region), Raipur Government Hatchery (RH), Luxmipur (Eastern region) and Sonali Hatchery (SH), Jessore (South-West region) (Fig. 1). Fingerlings approximately of the same size were collected in live condition directly from the hatcheries. Sampling sites in both the cases were selected in order to cover the genetic variation of a wide geographical range of the species in the country. For genetic analysis, 30 individuals from each source (population) were collected randomly and approximately 50 mg of fin tissue sample was clipped and used for DNA extraction.

We extracted genomic DNA from the fin tissue samples following the standard procedure of proteinase-K digestion, phenol: chloroform: isoamylalcohol extraction and alcohol precipitation as described by Islam and Alam [10]. The quality of DNA was checked by electrophoresis on 1% agarose gel and the quantity was ascertained using a spectrophotometer.

2.2 PCR Amplification

In absence of homologous primers for *C. cirrhosus*, we used five heterologous microsatellite primer sets developed from *Cyprinus carpio* [17] and *Barbodes gonionotus* [18] identified by Lal et al. [14]. Polymerase chain reaction contained 50 ng template DNA, 0.25 µM of each primer, 0.25 mM each dNTPs, 1 unit Taq DNA polymerase 1XPCR buffer containing 1.5 mM $MgCl_2$ in a total volume of 10 µl. The temperature profile consisted of 3 min initial denaturation at 94ºC followed by 35 cycles of 30 sec at 95ºC, 30 sec at 55ºC, 1 min at 72ºC ending with 5 min at 72ºC. We used a Master Cycler Gradient thermocycler of Eppendorf, Germany for conducting PCR.

2.3 Electrophoresis of PCR Product

For checking quality of amplification, the PCR products (10 µl) were mixed with 2.5 µl loading buffer (0.25% bromophenol blue, 0.25% xylene cyanol and 30% glycerol) and half of the mixture was run on 2% standard agarose gel containing ethidium bromide. The PCR products showing good resolution on agarose gel were then separated on 6% denaturing polyacrylamide gel containing 19:1 acrylamide:bis-acrylamide and 7M urea using a Sequi Gen GT sequencing electrophoresis system (BIO-RAD Laboratories, Hercules, CA). The silver nitrate staining protocol of Promega (Madison, WI) was applied for visualization of the alleles on the gel plate.

2.4 Scoring and Analysis of Microsatellite Data

Genotype of each individual fish was determined and recorded from the silver nitrate-stained gels for each microsatellite locus. Bands representing particular alleles at the microsatellite loci were scored from the plate manually and the size of the bands was estimated using the DNAfrag program version 3.03 [19]. A single matrix of genotypic data comprising the five microsatellite loci and six populations were constructed in Microsoft Excel. The genotype data were examined for any evidence of large allele dropouts or stutter-bands at the five loci using the software MicroChecker [20]. For further analysis, only data for the polymorphic loci were used. The polymorphic information content of the marker were calculated using the software CERVUS [21]. Estimation of allele frequencies was performed for each of the six populations of *C. cirrhosus* using GenAlex program version 6.4

[22]. The allelic richness (Ar), observed heterozygosity (H_O) and expected heterozygosity (H_E) were estimated by the software FSTAT 2.9.3.2 [23]. Markov chain method was used for calculating the exact P-value for testing the deviation from Hardy-Weinberg expectation at each locus implemented by the software GENEPOP 4.0 [24] with the following parameters: Dememorization-1000, batches- 100 and iterations per batch- 1000. The same software was used for exact G-test for homogeneity among the allelic distribution at different loci in the population pairs. The pairwise population differentiation (F_{ST}) values were calculated by means of 10,000 permutations of genotypes using the software FSTAT [23]. Wherever appropriate, the global α-value (P=0.05) was adjusted following Bonferroni and sequential Bonferroni correction [25]. A dendrogram was constructed using the Unweighted Pair Group Method of Averages (UPGMA) of Sneath and Sokal [26] implemented by the software MEGA5 [27].

3. RESULTS AND DISCUSSION

3.1 Within Population Genetic Variation

In the present study we have characterized five heterologous microsatellite DNA markers- three developed from common carp and two developed from silver barbs in *C. cirrhosis* collected from three rivers and three hatcheries. Examination of genotyping errors using MicroChecker revealed no evidence for large allele dropout or stutter-band scoring at any of the five loci. However, null alleles were detected in one locus and were corrected before analyzing the data as per van Oosterhaut et al. [20]. Out of the five, three (60%) were found to be polymorphic. The strength of a molecular marker for analyzing genetic diversity is evaluated on the basis of its polymorphic information content (PIC) estimated from the number and frequency of alleles. Botstein et al. [28] classified markers with PIC>0.5 as highly informative, 0.25<PIC<0.5 as middle informative and PIC<0.25 is slightly informative. In our study, of the three polymorphic markers analyzed in mrigal, two are highly informative and one is middle informative (Table 1).

A total of 10 alleles were detected at the five loci analyzed (Table 1). The number of allele per locus ranged from 1-3 with a mean of 2.000 alleles per locus. Three loci were found to be polymorphic in all the populations: MFW2 and

MFW17 with three alleles each and Bgon22 with two alleles. The mean allelic richness was the same (2.666) in all the populations. Lal et al. [15] detected 3-4, 2-4 and 2-3 alleles at locus MFW2, MFW17 and Bgon22 respectively in five populations of *C. cirrhosus* in India. On the other hand Chauhan et al. [29] detected 5, 6, and 3 alleles at the locus MFW2, MFW17 and Bgon2 in 10 river populations of *C. cirrhosus* in India. Kamonrat et al. [18] detected two alleles at the locus Bgon22 in *Barbodes gonionotus* which together with our observation indicate that variability at Bgon22 as a microsatellite locus is relatively low.

The allelic richness ranged from 2.000 to 3.000; the observed heterozygosity (H_O) ranged from 0.233 (BH, locus MFW17) to 0.633 (BH and RH), the expected heterozygosity (H_E) ranged from 0.406 (JR, locus, MFW2) to 0.664 (the HR, MFW2) and the F_{IS} values ranged from 0.032 (the Padma, locus MFW2) to 0.635 (BH, locus MFW17). The average H_O in the RH stock (0.577) was the highest and that of the JR population (0.355) was the lowest (Table 2). The H_E of the SH stock (0.602) was the highest and that of the JR population (0.506) was the lowest. The F_{IS} value of the PR population was the lowest and that of the BH stock was the highest (Table 2). The mean H_O and H_E values of the hatchery stocks were slightly higher than those of the river populations. Within the three wild populations the H_O was the highest (0.533) in the PR population and the lowest (0.355) in the JR population. The average H_O found in the present study was similar to that reported by Lal et al. [15]. These authors recorded the highest value of 0.333 in the Satluj river and the lowest value of 0.247 in the Ganga river population.

In all the cases the F_{IS} values were positive which means that all the six populations were deficient in heterozygosity. We detected significant deviations from Hardy-Weinberg expectations (HWE) in six out of 18 tests after Bonferroni corrections (Table 2). The SH stock deviated at two loci while the HR, JR, PR and the BH populations deviated at one locus each. No nonconformity to HWE was detected in the RH stock.

Among the six stocks of the present study, although the proportion of polymorphic loci was the same for all stocks yet the higher average values of H_O and H_E indicates that the genetic variability in the RH stock is higher than the other stocks. Among the three river populations, the genetic variability in the PR population was higher than the other two river populations. The higher diversity shown by the samples of the PR population was probably due to the fact that it is a large river (120 km long, 4 to 8 km wide) offering a wide gene pool. Generally, interbreeding with hatchery populations could reduce genetic variability of a natural stock [30]. This might have happened in case of the Padma

and the Jamuna populations but this may not be the case for loss of genetic variability of the population from the Halda river. Recently it has been reported that the population size of broodfish in the Halda river is reducing because of year round catching of brood fish and intensive collection of fertilized eggs by nearby people [31].

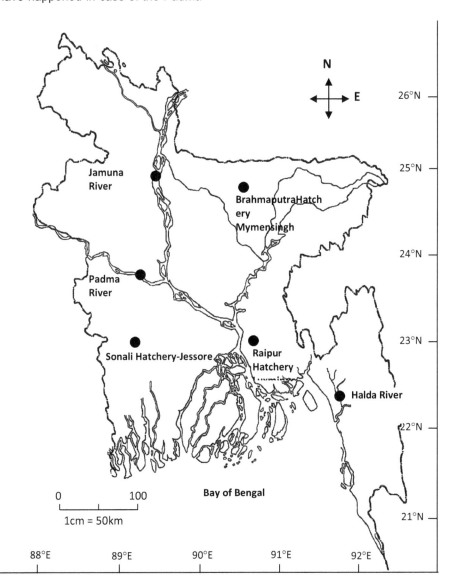

Fig. 1. Map of Bangladesh showing the six sites (•) from where the samples of *C. cirrhosus* samples were collected

Table 1. Allele frequency at five loci of *C. cirrhosus*. N is the number of individuals scored; PIC: Polymorphic Information Content; M: Monomorphic

Locus	PIC	Allele (bp)	Populations					
			HR	JR	PR	BH	RH	SH
			N=30	N=30	N=30	N=30	N=30	N=30
MFW1	M	164	1.000	1.000	1.000	1.000	1.000	1.000
MFW2	0.5	175	0.333	0.133	0.200	0.350	0.333	0.417
	85	172	0.367	0.117	0.300	0.383	0.400	0.233
		170	0.300	0.750	0.500	0.267	0.267	0.350
MFW17	0.5	225	0.250	0.183	0.217	0.283	0.383	0.187
	81	220	0.350	0.383	0.183	0.467	0.350	0.467
		216	0.400	0.433	0.600	0.250	0.267	0.350
Bgon22	0.3	113	0.700	0.400	0.500	0.417	0.683	0.383
	73	110	0.300	0.600	0.500	0.583	0.317	0.617
Barb54	M	97	1.000	1.000	1.000	1.000	1.000	1.000

Table 2. Genetic variation in *C. cirrhosus* populations at three microsatellite loci [allelic richness (*Ar*), Heterozygosity observed (H_O), Heterozygosity expected (H_E), Fixation Index (Inbreeding coefficient) (F_{IS}) and probabilities of conformity to Hardy-Weinberg equilibrium (HWEP)]

Locus	Parameters	Population (River)				Population (Hatchery)			
		HR	JR	PR	Mean	BH	RH	SH	Mean
MFW2	Ar	3.000	3.000	3.000	**3.000**	3.000	3.000	3.000	**3.000**
	H_O	0.433	0.433	0.600	**0.488**	0.533	0.600	0.300	**0.477**
	H_E	0.664	0.406	0.620	**0.563**	0.659	0.658	0.649	**0.655**
	F_{IS}	0.348	0.067	0.032	**0.149**	0.191	0.088	0.538	**0.272**
	HWEP	0.000**	0.365	0.025		0.181	0.784	0.000**	
MFW17	Ar	3.000	3.000	3.000	**3.000**	3.000	3.000	3.000	**3.000**
	H_O	0.500	0.300	0.467	**0.422**	0.233	0.500	0.333	**0.355**
	H_E	0.655	0.632	0.559	**0.615**	0.639	0.659	0.626	**0.641**
	F_{IS}	0.237	0.525	0.166	**0.309**	0.635	0.242	0.468	**0.448**
	HWEP	0.213	0.001**	0.004*		0.000**	0.185	0.001*	
Bgon22	Ar	2.000	2.000	2.000	**2.000**	2.000	2.000	2.000	**2.000**
	H_O	0.433	0.333	0.533	**0.433**	0.633	0.633	0.533	**0.599**
	H_E	0.406	0.480	0.500	**0.462**	0.486	0.455	0.491	**0.477**
	Fis	0.067	0.306	0.067	**0.146**	0.303	0.392	0.086	**0.260**
	HWEP	1.000	0.128	1.000		0.147	0.052	1.000	
Population means over all loci									
Ar		2.666	2.666	2.666	**2.666**	2.666	2.666	2.666	**2.666**
H_O		0.455	0.355	0.533	**0.447**	0.466	0.577	0.388	**0.477**
H_E		0.575	0.506	0.559	**0.546**	0.594	0.591	0.602	**0.591**
F_{IS}		0.217	0.299	0.088	**0.201**	0.376	0.241	0.364	**0.327**

*Significant at P<0.0167), ** P<0.0033) after Bonferroni corrections [25]*

3.2 Inter-population Genetic Variation

We compared the allele frequencies among the population pairs at three polymorphic loci. Significant non-homogeneity in allele frequencies in pairs of populations was detected in nine cases out of 45 comparisons. The HR-JR pair showed non-homogeneity at two loci while the HR-BH, JR-BH, JR-RH, JR-SH, PR-BH, PR-RH and the PR-SH population pairs showed non-homogeneity at one locus each (Table 3).

Population pair-wise comparisons of the overall F_{ST} values detected four significant differentiations when tested on the basis of sequential Bonferroni correction (Table 4). Significant population structuring was observed between the HR-JR, JR-BH, JR-RH and the PR-RH population pairs (Table 4). The low genetic diversity in two hatchery populations except the RH population seems to indicate that the hatchery practices might have led to the loss of genetic variation. In case of *C. catla* and *Labeo rohita*, the genetic variations, both in terms of expected heterozygosity and allelic richness were reported to be lower in hatchery samples compared to the river samples [6,7]. Among the river populations, significant F_{ST} value was calculated only between the HR and the JR population that could be explained by the geographical barrier of the two river systems. In case of hatchery populations, no significant F_{ST} value was observed in any of the population-pairs. However, the values in the BH-JR, RH-JR and RH-PR population pairs were found to be significant. Hansen et al. [6] and Khan et al. [7] also detected significant structuring between the river and hatchery populations of catla and rohu respectively.

3.3 Genetic Distance and Dendrogram

A matrix of genetic distance [32] was built based on allelic frequencies of polymorphic loci (Table 5). The highest genetic distance value (D=0.221) was measured between the RH and the JR populations while the lowest genetic distance value (D=0.019) was estimated between the RH and the HR populations. Among the river populations the genetic distance between the JR and the HR population (0.203) was estimated to be larger than that between the PR-JR (0.068) and the PR–HR (0.091) population pairs. In case of hatchery populations, the genetic distance value (0.049) between the RH and BH was larger

than that observed between the SH-BH (0.022) and the RH-SH population pairs.

The UPGMA dendrogram based on the Nei's [32] genetic distance among the population pairs resulted in two major clusters: the Halda river and RH populations were in one cluster and the remaining four populations were in the other cluster. The second cluster was further separated into two sub-clusters: the JR and PR populations were in one group and the BH and SH were in the other group (Fig. 2). The higher genetic distance between the HR and the JR populations may be due to the geographical distance between the Halda and the Jamuna rivers. But the lower genetic distance between the HR and the PR population which was similar to the genetic distance between JR and PR populations was surprising as the Halda river is located in the eastern part of the country having no direct connection with the Padma river. The lack of differentiation between HR and PR populations might be due to the low sample size. The clustering of the HR and the RH population indicates that the origin of the brood fish of Raipur Hatchery is the Halda river. The grouping between the JR and the PR population was expected because the Jamuna and the Padma rivers join each other near the Goalandaupazila of Rajbari district, making a high possibility of mixing between their fish populations.

Table 3. Exact G-test for differences in allele frequencies at different locus among the population-pairs

Population pairs	P-values at different loci		
	MFW2	MFW17	Bgon22
HR-JR	0.0000**	0.6942	0.0008**
HR-PR	0.0794	0.0613	0.0238
HR-BH	0.9472	0.1967	0.0015**
HR-RH	0.9239	0.2006	0.5561
HR-SH	0.2962	0.4241	0.1257
JR-PR	0.0169	0.0509	0.3610
JR-BH	0.0000**	0.0974	1.0000
JR-RH	0.0000**	0.0326	0.0109
JR-SH	0.0000**	0.5698	0.1013
PR-BH	0.0238	0.0001**	0.4523
PR-RH	0.0271	0.0015**	0.1383
PR-SH	0.0394	0.0033**	0.5919
BH-RH	1.0000	0.3888	0.0164
BH-SH	0.2427	0.3280	0.1416
RH-SH	0.1657	0.0564	0.4595

*** Significant after sequential Bonferroni correction [25]*

Table 4. MultilocusF_{ST} (above diagonal) and P-values (below diagonal) between the pairs of six
C. cirrhosus populations across all loci

	HR	JR	PR	H	H	H
HR	-	0.1178*	0.0437NS	0.0390NS	0.000NS	0.045NS
JR	0.0033*	-	0.0370NS	0.0894*	0.1258*	0.0689NS
PR	0.0267NS	0.0233NS	-	0.0629NS	0.0620*	0.0423NS
BH	0.0767NS	0.0033*	0.0067NS	-	0.0175NS	0.0057NS
RH	0.5800NS	0.0033*	0.0033*	0.2267NS	-	0.0121NS
SH	0.4100NS	0.0067NS	0.0167NS	0.3033NS	0.1267NS	-

* Significant after sequential Bonferroni correction [25]

Table 5. Summary of Nei's [29] genetic genetic distance (below diagonal) values between pairs
of six populations of _C. cirrhosus_ based on three loci

	HR	JR	PR	BH	RH	SH
HR	***					
JR	0.203	***				
PR	0.091	0.068	***			
BH	0.092	0.157	0.127	***		
RH	0.019	0.221	0.122	0.049	***	
SH	0.038	0.123	0.092	0.022	0.036	***

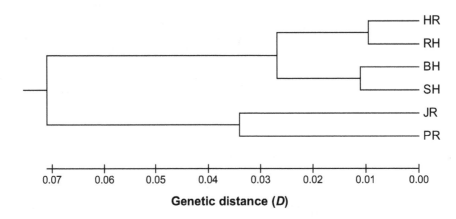

Genetic distance (_D_)

Fig. 2. UPGMA dendrogram based on Nei's [32] genetic distance (_D_) estimated from allele
frequencies at three microsatellite loci of three river and three hatchery samples of
C. cirrhosus
HR-Halda River; JR-Jamuna River; PR-Padma River; BH-Brahmaputra Hatchery; RH-Raipur Hatchery; SH-
Sonali Hatchery

4. CONCLUSION

We have analyzed the genetic structure of three river and three hatchery populations of the Indian major carp _C. cirrhosus_ using heterologous microsatellite markers developed from other carp species. We have detected a relatively low level of genetic variation both in the river and hatchery populations in terms of allelic richness and heterozygote deficiency. As the nonhomologous markers were found to be less variable, it is essential to develop microsatellite markers from the genome of _C. cirrhosus_ for a more detailed study about the genetic population structure of this species.

ETHICAL APPROVAL

All authors hereby declare that all experiments have been examined and approved by the appropriate ethics committee and have therefore been performed in accordance with the ethical standards laid down in the 1964 Declaration of Helsinki.

COMPETING INTERESTS

Authors have declared that no competing interests exist.

REFERENCES

1. FAO. FAO Yearbook, Fishery and Aquaculture Statistics 2010. Food and Agriculture Organization of the United Nations, Rome, Italy. 2012;78.
2. Rajts F, Huntingdon T, Hussain MG. Carp brood stock management and genetic improvement programme under Fourth Fisheries Project. In: Penman DJ, Hussain MG, McAndrew BJ, Mazid MA. (eds.). Proceedings of a workshop on Genetic Management and Improvement Strategies for Exotic Carps in Asia, Dhaka, Bangladesh. 2002;95-106.
3. DoF. Annual Report. Department of Fisheries, Ministry of Fisheries and Livestock Bangladesh. 2011;76.
4. Alam MA, Akanda MSH, Khan MMR, Alam MS. Comparison of genetic variability between a hatchery and a river population of rohu (*L. rohita*) by allozyme electrophoresis. Pak J Biol Sci. 2002;5(9):959-61.
5. Simonsen V, Hansen MM, Mensberg KLD, Sarder MRI, Alam MS. Widespread hybridization among species of Indian major carps in hatcheries, but not in the wild. J Fish Biol. 2005;67:794-808.
6. Hansen MM, Simonsen V, Mensberg KLD, Sarder MRI, Alam MS. Loss of genetic variation in hatchery-reared Indian major carp, *Catla catla*. J Fish Biol. 2006;69(B):229-41.
7. Khan MMR, Alam MS, Bhuiyan MMH. Allozyme variation of hatchery and river populations of rohu (*Labeo rohita* Hamilton) in Bangladesh. Aquac Res. 2006;37:233–40.
8. Vandewoestijne S, Schtickzelle N, Baguette M. Positive correlation between genetic diversity and fitness in a large, well-connected metapopulation. BMC Biol. 2008;6:46.
9. Alam MS, Islam MS. Population genetic structure of *Catla catla* (Hamilton) revealed by microsatellite DNA markers. Aquaculture. 2005;246:151-60.
10. Islam MS, Alam MS. Randomly Amplified Polymorphic DNA analysis of four different populations of the Indian major carp *Labeo rohita* (Hamilton). J ApplIchthyol. 2004;20:407-12.
11. Alam MS, Jahan M, Hossain MM, Islam MS. Population genetic structure of three major river populations of Rohu, *Labeo rohita* (Cyprinidae: Cypriniformes) using microsatellite DNA markers. Genes Genom. 2009;31(1):43-51.
12. Hasanat MA, Mollah MFA, MS Alam. Assessment of genetic diversity in wild and hatchery populations of mrigal *Cirrhinus cirrhosus* (Hamilton-Buchanan) using allozyme markers. Int J Fish Aquat Stud. 2014;1(4):24-31.
13. McConnell SK, Hamilton L, Morris D, Cook D, Pacquet D, Bentzen P, Wright JM. Isolation of salmonid microsatellite loci and their application to the population genetics of Canadian east coast stocks of Atlantic salmon. Aquaculture. 1995;137:19-30.
14. Reilly A, Elliott NG, Grewe PM, Clabby C, Powell R, Ward RD. Genetic differentiation between Tasmanian cultured Atlantic salmon (*Salmo salar* L.) and their ancestral Canadian population: Comparison of microsatellite DNA and allozyme and mitochondrial DNA variation. Aquaculture. 1999;173:459-69.
15. Lal KK, Chauhan T, Mandal A, Singh RK, Khulbe L, Ponniah AG, Mohindra V. Identification of microsatellite DNA markers for population structure analysis of Indian major carp, *Cirrhinus mrigala* (Hamilton-Buchanan, 1822). J ApplIchthyol. 2004;20:87-91.
16. Swain S, Das SP, Bej D, Patel A, Jayasankar P, Chaudhary BK, etc. Evaluation of genetic variation in *Labeo fimbriatus* (Bloch, 1795) populations using heterologous primers. Indian J Fish. 2013;60(1):29-35.
17. Crooijmans RPMA, Bierbooms VAF, Komen J, Poel JJV, Groenen MAM. Microsatellite markers in common carp (*Cyprinus carpio* L.). Anim Genet. 1997;28:129-34.
18. Kamonrat W, McConnell SK, Cook JDI. Polymorphic microsatellite loci from the Southeast Asian cyprinid, *Barbodes gonionotus* (Bleeker). MolEcol Notes. 2002; 2: 89–90.
19. Nash JHE. DNAfrag, Version 3.03. Institute for Biological Sciences, National Research Council of Canada, Ottawa, Ontario, Canada; 1991.
20. Van Oosterhout CV, Hutchinson WF, Wills DPM, Shipley P. MICRO- CHECKER:

software for identifying and correcting genotyping errors in microsatellite data. Mol Ecol Notes. 2004;4:535–8.

21. Kalinowski ST, Taper ML, Marshall TC. Revising how the computer program CERVUS accommodates genotyping error increases success in paternity assignment. Mol Ecol. 2007;16(5):1099–106.

22. Peakall R, Smouse PE. GenAlEx V6.4: Genetic Analysis in Excel. Population genetic software for teaching and research. Mol Ecol Notes. 2006;6:288-95.

23. Goudet J. FSTAT Version 1.2: a computer program to calculate F-statistics. J Hered. 1995;86:485-86. FSTAT 2.9.3.2 [Software] Available:http://www2.unil.ch/popgen/softwares/fstat.htm

24. Rousset F. GENEPOP'007: A complete re-implementation of the GENEPOP software for Windows and Linux. Mol Ecol Resour. 2008;8:103-6.

25. Rice WR. Analysing tables of statistical tests. Evolution. 1989;43:223-5.

26. Sneath PHA, Sokal RR. Numerical Taxonomy. Freeman, San Francisco; 1973.

27. Tamura K, Peterson D, Peterson N, Stecher G, Nei M, Kumar S. MEGA5: Molecular Evolutionary Genetics Analysis using Maximum Likelihood, Evolutionary Distance, and Maximum Parsimony Methods. Mol Biol Evol. 2011; 28: 2731-9.

28. Botstein D, White RL, Skolnick M, David R. Construction of a genetic linkage map in man using restriction fragment length polymorphisms. Am J Hum Genet. 1980;32: 314-31.

29. Chauhan T, Lal KK, Mohindra V, Singh RK, Punia P, Gopalakrishnan A, etc. Evaluating genetic differentiation in wild populations of the Indian major carp, *Cirrhinus mrigala* (Hamilton–Buchanan, 1882): Evidence from allozyme and microsatellite markers. Aquaculture. 2007;269:135-49.

30. Hindar K, Ryman N, Utter F. Genetic effects of cultured fish on natural fish populations. Can J Fish Aquat Sci. 1991;48:945-57.

31. Azadi MA, Arshad-ul-Alam M. Spawn fishery of major carps (*Catla catla, Labeo rohita, Cirrhinus cirrhosus* and *Labeo calbasu*) of the river Halda. Book of Abstract. Fifth Fisheries Conference and Research Fair. Dhaka Bangladesh; 2012.

32. Nei M. Genetic distance between populations. Am Nat. 1972;106:283-92.

Establishment of a Simple Plant Regeneration System Using Callus from Apomictic and Sexual Seeds of Guinea Grass (*Panicum maximum*)

Chen Lanzhuang[1*], Nishimura Yoshiko[1], Umeki Kazuma[1], Zhang Jun[2] and Xu Chengti[2]

[1]*Faculty of Environmental and Horticultural Science, Minami Kyushu University, 3764-1, Tatenocho, Miyakonojo City, Miyazaki, 885-0035, Japan.*
[2]*Qinghai Academy of Animal Science and Veterinary Medicine, Xining, Qinghai, 810016, China.*

Authors' contributions

This work was carried out in collaboration by the authors. Authors CL, ZJ and XC designed the study and author CL wrote the first draft of the manuscript. Authors NY and UK performed the experiments and data management. Authors ZJ and XC managed the literature searches. All authors read and approved the final manuscript.

Editor(s):
(1) Standardi Alvaro, Department of Agricultural and Environmental Sciences, University of Perugia, Italy.
Reviewers:
(1) Anonymous, Universidade Federal de Uberlândia, Brazil.
(2) Paula Cristina da Silva Angelo, Embrapa Western Amazon, Brazil.
(3) Anonymous, China Agricultural University, China.
(4) Anonymous, University of Agriculture in Krakow, Poland.
(5) Anonymous, Lille University of Science and Technology, France.

ABSTRACT

Aims: For analysis of apomixis genes, as the first step, an efficient and simple plant regeneration system has been established using callus from apomictic and sexual seeds of guinea grass (*Panicum maximum*).
Study Design: The best basic medium for callus formation from matured seeds of guinea grass was selected, and then, the best combinations of growth regulators on different media were selected for plant regeneration by indirect organogenesis.
Methodology: Guinea grass accessions of sexual and apomict of seeds were used for culture. Seeds sterilized were cultured in Murashige and Skoog [10] medium (MS) and in that proposed by Chu et al. [12] (N6D) for callus formation. The best medium and, the effects of L-proline and growth

regulators' type and concentrations on callus formation and plant regeneration in 3 accessions were examined, respectively. After the plantlets rooting in MS hormone-free medium, the complete plants were planted in pots for hardening.

Results: N6D medium has given better rates for callus formation, that is, 97.1% in sexual N68/96-8-o-5, and 91.7% and 84.6% in apomict N68/96-8-o-11 and 'Natsukaze', respectively. MS medium with 0.2 mg/l of Kinetin has given the best rate or plant regeneration among the used 4 kinds of the media. For each material, the best results were obtained on MS with 0.2 mg/l of Kinetin for N68/96-8-o-5, and MS with 2.0 mg/l of L-proline and 0.2 mg/l of Kinetin for 'Natsukaze'. After hardening of the regenerated plants in soil, 100% of surviving rates were obtained, and showed normal growth comparing with the plants-derived from seeds.

Conclusion: We have established, as the first case, an efficient and simple plant regeneration system by using callus from apomictic and sexual seeds of guinea grass for analysis of apomixis genes, consisting of analysis of the best media, L-proline usages and phytohormone combinations for callus formation, plant regeneration and hardening in different stages, respectively.

Keywords: Apomixis; apomixis specific gene; In vitro culture; culture medium; growth regulators; Panicum maximum; mature seeds.

1. INTRODUCTION

Guinea grass (*Panicum maximum*) is an important tropical forage crop. It represents in nature by both of tetraploid biotypes that can reproduce through apomixis, and sexual one [1]. Apomixis is a unique reproductive mode that by passes meiosis to produce seed genetically same to the mother parent [2]. So it is expected to promise the economic benefits over than the "Green Revolution" by using apomixis to fix F_1 plants as a variety in agriculture [3]. Up to now, we based on the results of cytological observation of embryo sac formation in both of sexual and apomictic guinea grass (*Panicum maximum*) by using Nomarski differential interference-contract microscopy (DIC) [4,5] and used differential screening method based on the ovary length as an index, and have isolated successfully the apomixis specific gene (*ASG-1*) [6] which specifically expressed in the stage of appearance of aposporous embryo sac initial cell (AIC) [7]. The application of apomixis as a technology to "clone" will be a good idea for plants that express genes homologous to the *ASG-1*, isolated from facultatively apomictic guinea grass [6], exerting the same function in similar tissues of embryo sac and surroundings. In order to do functional analysis of *ASG-1*, using gene transformation method mediated by *Agrobacterium*, the efficient and simple plant regeneration system must be established from the donor plant, that is, guinea grass. In addition, we have made the *ASG-1* transgenic plants in rice and *Arabidopsis* [8,9]. However, the systems from the most important and necessary plants of guinea grass have not been established in efficient plant regeneration and transformant using *Agrobacterium*. In this study, we focused on establishing system of plant regeneration from matured seeds of accessions of sexual and apomict and apomictic variety of guinea grass.

2. MATERIALS AND METHODS

2.1 Plant Materials

Guinea grass (*Panicum maximum* Jacq.) accessions of sexual N68/96-8-o-5, and apomict of N68/96-8-o-11 from which the *ASG-1* was isolated, and apomictic variety of "Natsukaze", cultured in Field Center of Minami Kyushu University, were used as materials for seed donor, respectively. The seeds were collected from living plants, dried for about one week in desiccator, and used for culture. This study was carried out in Faculty of Environmental and Horticultural Science, Minami Kyushu University, between April 2013 and October 2014. For seed sowing, at first, the matured seeds of 5g were put into the 50 ml tube containing 2.5% concentration of sodium hypochlorite solution, and 0.1% of tween 20, and with shaking in 120 rpm at room temperature for 3 hours for sterilization (Fig. 1A). And then, the tube was moved into biohazard and washed with sterilized water up to the bubbles faded away. The seeds were put onto the filter paper in the dish, and dried by using tweezer to take off the palea (Fig. 1B). As the preliminary examination, when the matured seeds just harvested were sowed immediately into the callus formation media, the germination of the seeds was not uniform. According to the phenomenon, if the seeds were used for gene transformation, it is worried that the transgenic plants would also be not uniform.

When we used the seeds having already been harvested for over 3 months kept in laboratory room, the uniform germination was obtained from the seeds. From this result, it is considered that the seeds of guinea grass need breaking of dormancy, so that we used the seeds over 3 months keeping in room temperature for giving uniform germination.

2.2 Culture Media for Callus Formation

The seeds were put into the dishes, respectively containing the Murashige and Skoog (MS) medium [10] with 30,000 mg/l of sucrose, 5 mg/l of 2,4-D, 3,200 mg/l Gellan Gum, pH 5.8 [11] and Chu et al. [12] (N6D) medium with 30,000 mg/l of sucrose, 5 mg/l of 2,4-D, 3,200 mg/l Gellan Gum, pH 5.8 [11], and N6D medium [12] with 30,000 mg/l of sucrose, 3,981 mg/l of CHU powder [12], 300 mg/l of casamino acid, 2,878 mg/l of L-proline, N6D vitamins, 2 mg/l of 2,4-D, 8,000 mg/l of agar, pH5.8. The culture was kept in 30°C, dark for callus formation in interval subculture of 3 weeks.

After one month of culture of seeds, the rates of callus formation and the difference in the rates among sexual of N68/96-8-o-5, and apomicts of N68/96-8-o-11 and 'Natsukaze' were examined on the N6D medium. For each accession, 100 seeds were used for culture and the same experiment was repeated for three times.

2.3 Culture Media for Plant Regeneration

In order to find out the best medium for plant regeneration, the calli derived from matured seeds of N68/96-8-o-11 were cultured on different regeneration media. As there were not proper medium reported previously for guinea grass, the examination of the efficient condition was carried out on four kinds of media [13,11,14,15] (Table 1), which were used for plant regeneration from different tissues (not seeds) of monocotyledon plants.

For the calli of N68/96-8-o-5, N68/96-8-o-11 and 'Natsukaze', MS medium with 30g/l of sucrose, 0.2 mg/l Kinetin, pH5.8, and 8 g/l agar was used to examine the regeneration rates. In addition, an essential amino acid of L-proline (2 mg/l) was added into MS medium and used for the 3 same materials. As in the pre-treatment, different effects were observed in the 3 kinds of materials, the best combination of growth regulator types was also examined. The regeneration rates were investigated in MS medium with 2 mg/l of Kinetin, and 0.1, 0.2 and 0.3 mg/l of naphthalene acetic acid (NAA), respectively. For each accession, 30 calli were cultured and the same experiment was repeated for three times.

Fig. 1. The matured seeds of guinea grass treated for germination. A: seed sterilization by shaking machine; B: seeds in the dish after washing by sterile water

Table 1. Comparison of the best combination of growth regulators reported previously in different materials used in this study for plant regeneration from callus derived from matured seeds of guinea grass*

Kinds of materials	NAA	GA₃	Kinetin	BAP
Rice [13]	0.02 mg/l	-	2.0 mg/l	-
Guinea grass [11]	0.01 mg/l	-	2.0 mg/l	-
Switch grass [14]	0.2 mg/l	0.5 mg/l	-	1.0 mg/l
Toll fescue [15]	-	-	0.2 mg/l	-

Murashige & Skoog [10] medium was used as the basic medium

2.4 Hardening of Regenerated Plantlets

After the plantlets were cultured in MS with growth regulators free medium for 2-3 weeks, the complete plants with well-developed roots were planted in pot containing metro mix (Hyponex, Japan) and vermiculite 1:1 (v:v) for hardening under the conditions of 30°C, 24 hours of day light. The same experiment was repeated for three times.

3. RESULTS AND DISCUSSION

3.1 Evaluation of the Culture Media for Callus Formation from Matured Seeds

For the examination of media on callus formation in guinea grass, Chen et al. [11] have used the basic MS medium complemented with 5~10 mg/l of 2, 4-D for the leaflet culture. In this study, the same medium gave the slow growth of callus compared with that of N6D, even though the callus formed as same in both media (Fig. 2A, B). In the N6D which gave higher rates of callus formation, L-proline was added with 2.878 mg/l. Li & Qu [14] also indicated that the addition of L-proline is efficient to callus formation used for the transformation of switch grass. Therefore, the addition of L-proline and N6D medium which is well used in rice culture were adopted in this study for callus formation from matured seeds of guinea grass.

3.2 Difference in Callus Formation Rates between Sexual and Apomict

Chen et al. [11] has reported that there have differences between the accessions and between sexual and apomict in cultures of leaflets of guinea grass. In the culture of matured seeds there also appeared differently in callus formation (Figs. 3A, B, C). In sexual N68/96-8-o-5 accession, while the seed germinated designated as in Fig. 3D, the callus located in the root pole, derived from embryo formed, and later, the callus showed vigorous growth in sizes of the big callus and small young leaflet. However, in apomicts of N68/96-8-o-11 and 'Natsukaze', while the seeds germinated designated as in Figs. 3B and C, the shoots grew firstly and fast, and then, callus formed in the root pole, after the transition from shoot to callus. And as a contrast to the sexual, the callus of apomict showed the sizes of small callus and big young leaflet. The difference of callus formation appeared between sexual and apomict may be considered as the difference of

growth regulators' balance existed in both of them.

In the culture of matured seeds there also appeared differently in callus formation (Fig. 3A, B, C). In sexual N68/96-8-o-5 accession, while the seed germinated the callus derived from embryo formed, and later, the callus showed vigorous growth. However, in apomicts of N68/96-8-o-11 and 'Natsukaze', the shoot grew firstly, and then, callus formed from around root and young leaflet. The calli formed in all the 3 accessions showed white color and soft characteristics, excepting the apomict 'Natsukaze' which showed compact one in part of calli (Data not showed). However, all the 3 accessions gave high rates of callus formation after one month of culture. In particular, the sexual N68/96-8-o-5 gave the highest rate of 97.1% (Table 2), comparing with those of 91.7% and 84.6% in apomicts of N68/96-8-o-11 and 'Natsukaze', respectively. In this study, the fact that the sexual one showed vigorous growth of callus (Fig. 3D) and highest rate, can be considered as the important and positive factor to the further transformation experiment of *ASG-1* for its functional analysis, even though the reasons were not yet clear.

Table 2. The rates of callus formation from matured seeds in N6D medium

Accessions	Rates of callus formed (%)
N68/96-8-o-5	97.1±9.2 (165/170)[z]
N68/96-8-o-11	91.7±12.5 (266/290)
'Natsukaze'	84.6±11.9 (237/280)

[z](Number of callus formed/number of seeds cultured)

3.3 Examination of Callus Culture Media for Plant Regeneration

Fig. 3 shows callus formation in different accessions in the same medium, and showed the difference in callus formation after one and two months of culture. Fig. 3D showed calli after two months of culture in which, seedlings were disappeared gradually. However, when the calli were moved onto medium for plant regeneration, only calli without seedlings were used (Fig. 4).

Table 3 showed the rates of plant regeneration in different media from calli-derived from matured seeds of guinea grass. There appeared with brown color in callus cultured in the Rice medium [13], a well-used one in rice, and no color change and no growth in callus in the medium used in switch grass[14]. And more, there appeared with only root formation in RE medium [11] used in guinea grass. However, the shoot was regenerated in the medium used in tall fescue [15]. Therefore, it is considered that MS complemented with 0.2 mg/l of kinetin is the best selection for plant regeneration from the calli-derived from matured seeds of guinea grass (Figs. 4A, B, C and D).

3.4 The Effects of L-proline and, Growth Regulator Type and Concentration on Regeneration Rates of 3 Kinds of Guinea Grass

Table 3 shows the results of effects of growth regulators and L-proline on the rates of plant regeneration. When L-proline was not used in the same MS medium, the lowest rate of plant regeneration was 25.0% in apomict of N68/96-8-

o-11, and the highest one was 80.0% in sexual N68/96-8-o-5. And the same apomict of 'Natsukaze' gave the rate of 61.9% higher than that of N68/96-8-o-11. From those results, it is clear that the rates of plant regeneration are different among the accessions and variety used in this study.

To make it clear further that how L-proline effects the regeneration rate, 2.0 mg/l of L-proline (Table 3) was added for regeneration on MS medium. As a result, N68/96-8-o-5 showed no any regeneration with shoot and root, even though it gave the highest regeneration rate of 80.0% among the 3 accessions when the medium was added only with 0.2 mg/l of kinetin. However, the apomict of N68/96-8-o-11 gave the rate of 41.6% higher than the 25.0%, when the L-proline was not added. And more, the apomictic variety of

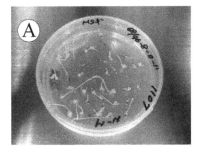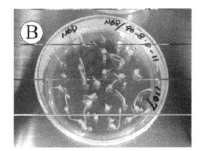

Fig. 2. Callus formation from the matured seeds of guinea grass cultured in different media. A: callus formed in MS medium; B: callus formed in N6D medium

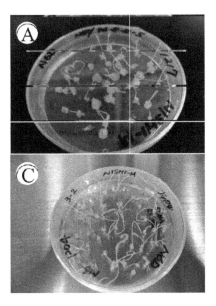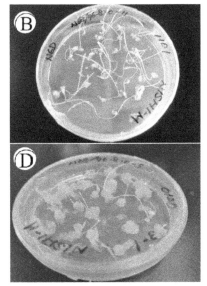

Fig. 3. Morphologies of calli formed in 3 accessions of guinea grass. A: calli in N68/96-8-o-5 after one month culture; B: calli in N68/96-8-o-11 after one month culture; C: calli in 'Natsukaze' after one month culture; D: calli in N68/96-8-o-5 after two months of culture

'Natsukaze' gave the highest rate of 91.7%. From the results obtained, L-proline may play a positive role for plant regeneration in apomicts but sexual one, even though the relation between L-proline and apomict was not clear. The materials of sexual and apomict used in this study were few, so that the further experiments are needed for understanding the effect of L-proline on the plant regeneration in apomicts.

As the above different results obtained from the different accessions, the best combinations of growth regulators including L-proline and NAA (Table 3) on MS medium for each accession was further investigated. Fig. 5 shows the plant regeneration from different combinations of growth regulators in different accession. There were not different morphologies observed among the different media. And from Table 3, it is clear for the best combinations of growth regulators on MS medium for each accession that, the best results of 80% in N68/96-8-o-5, 91.7% in 'Natsukaze', and 41.6% in N68/96-8-o-11 were obtained in 1) Kin:0.2 mg/l, 2)L-proline:2.0g/l and Kin:0.2 mg/l, and 3)L-proline:2.0 g/l, NAA:0.2 mg/l, and NAA:0.3 mg/l and Kin:0.2 mg/l, respectively. These results indicated that there existed in different regeneration rates among different materials and different combinations of growth regulators even on the same medium.

3.5 Hardening of Regenerated Guinea Grass

In general, a lot of regenerated plantlets need to be moved onto growth regulator free medium for rooting in direct or indirect organogenesis types, and then, for hardening. In this study, as the plantlets without rooting or with weak and few roots were obtained from callus derived from the matured seeds, it is clear that the type of plant regeneration should be indirect organogenesis. As described in M & M, the plantlets obtained were moved onto growth regulator free MS medium, and the whole plants with rooting were obtained in rates of 100% (Fig. 6A). As guinea grass is weak to high humidity condition, the regenerated plants with shoot and roots were transplanted into the 6 cm diameter of pot containing metro mix (Hyponex, Japan) and vermiculite (in 1:1 = v : v) for hardening under 30°C, 24 hours of day-length in a growth chamber. All of the plants obtained from all of media were subjected to acclimatization. As a result, all of the treated plants gave 100% of surviving rates, and showed normal growth comparing with the plants-derived from seeds (Fig. 6).

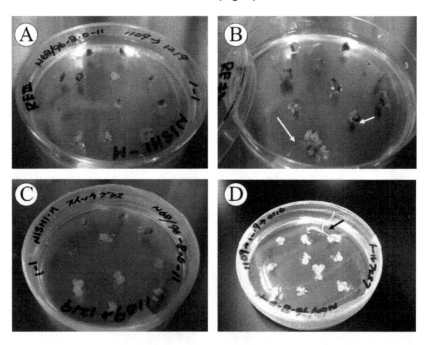

Fig. 4. Plant regeneration on 4 kinds of phytohormone combinations in MS media in different accessions. A: medium for rice[13]; B: medium for guinea grass[11]; C: medium for switch grass[14]; D: medium for toll fescue[15], referring to Table 1

Table 3. The effects of L-proline and NAA on the rates (%) of plant regeneration of guinea grass[z]

Accessions	Kin:0.2 mg/l	L-proline:2.0 g/l Kin:0.2 mg/l	NAA:0.1 mg/l Kin:0.2 mg/l	NAA:0.2 mg/l Kin:0.2 mg/l	NAA:0.3 mg/l Kin:0.2 mg/l
N68/96-8-o-5	80±10.4	0±0	8.3±3.4	16.6±3.6	0±0
N 68/96-8-o-11	25±6.4	41.6±5.7	8.3±2.5	41.6±8.8	41.6±7.9
'Natsukaze'	61.9±5.8	91.7±9.2	0±0	25.9±5.7	48.3±6.7

[z]Murashige & Skoog [10] medium was used as the basic medium; For each accession, 50 calli were cultured, respectively and the same was repeated 3 times

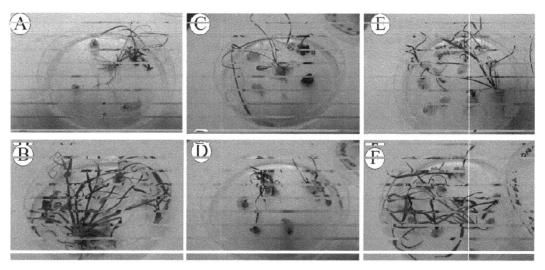

Fig. 5. Plant regeneration of guinea grass in different accessions and media. A & B: N68/96-8-0-5 (MS+0.2 mg/l : Kinetin); C & D: N68/96-8-0-11 (MS+0.2mg/l : NAA, 0.2 mg/l : Kinetin); E: 'Natsukaze' (MS+0.2 mg/l : Kinetin); F: 'Natsukaze' (MS+2.0g/l : L-proline+0.2mg/l : Kinetin)

Fig. 6. Hardening of regenerated plants of guinea grass. A: The regenerated plants before hardening; B: The plants after hardening; C: The whole plants are successfully hardened

4. CONCLUSION

We have established an efficient and simple plant regeneration system by using callus from apomictic and sexual seeds of guinea grass for analysis of apomixis genes, consisting of analysis of the best media, L-proline usages and growth regulator combinations for callus formation, plant regeneration and hardening in different stages, respectively. We are starting to do the *ASG-1* transformation with the established plant regeneration system.

ACKNOWLEDGEMENTS

This study was supported partly by a Grant-in-Aid from the Ministry of Education, Culture,

Sports, Science, and Technology of Japan (No. 23380009 to LZC).

COMPETING INTERESTS

Authors have declared that no competing interests exist.

REFERENCES

1. Savidan YH. Heredite de L'apomixie, contribution a L'etude de L'heredite de L'apomixie sue *Panicum maximum* Jacq. (analyse des embryonnaires). Cah ORSTOM Ser Biol. 1975;10:91-95.

2. Koltunow AM. Apomixis: Embryo sac and embryos formed without meiosis or fertilization in ovules. Plant Cell. 1993; 5:1425-1437.

3. Hanna WW, Bashaw EC. Apomxis: Its identification and use in plant breeding. Crop Science. 1987;27:1136-1139.

4. Chen LZ, Kozono T. Cytology and quantitative analysis of aposporous embryo sac development in guineagrass (*Panicum maximum* Jacq.). Cytologia. 1994a;59:253-260.

5. Chen LZ, Kozono T. Cytological evidence of seed-forming embryo development in polyembryonic ovules of facultatively apomictic guineagrass (*Panicum maximum* Jacq.). 1994b;59:351-359.

6. Chen LZ, Miyazaki C, Kojima A, Saito A, Adachi T. Isolation and characterization of a gene expressed during the period of aposporous embryo sac initial cell appearance in guineagrass (*Panicum maximum* Jacq.). Journal of Plant Physiology. 1999;154:55-62.

7. Chen LZ, Guan LM, Seo K, Hoffmann F, Adachi T. Developmental expression of *ASG-1* during gametogenesis in apomictic guineagrass (*Panicum maximum*). Journal of Plant Physiology. 2005;162:1141-1148.

8. Chen LZ, Nishimura Y, Tetsumura T, Hamaguchi T, Sugita T, Yoshida K, Xu C. Establishment of *Agrobacterium*-mediated transformation system in *Arabidopsis Thaliana* for functional analysis of *ASG-1*, an apomixis specific gene. International Journal of Current Biotechnology. 2013; 1(9):6-12.

9. Nishimura Y, Tetsumura T, Hamaguchi T, Sugita T, Ichikawa H, Yoshida K, Xu C, Chen LZ. Functional analysis of apomixis specific gene: Plant regeneration of transformed *Arabidopsis* with *ASG-1* gene using floral dip method. Bull. Minami Kyushu U. 2013;43(A):33-39.

10. Murashige T, Skoog F. A revised medium for rapid growth and bioassays with tobacco tissue culture. Physiol. Plant. 1962;15:473-479.

11. Chen LZ , Okabe R, Guan LM, Adachi T. A simple and efficient culture of leaflets for plant regeneration in guineagrass (*Panicum maximum*). Plant Biotechnology. 2002;19(1):63-68.

12. Chu CC, Wang CC, Sun CS, Hsu C, Yin KC, Chu CY, Bi FY. Establishment of an efficient medium for another culture of rice, through comparative experiments on the nitrogen sources. Sci. Sin. 1975;18:659-668.

13. Toki S, Hara N, Ono K, Onodera H, Tajiri A, Oka S, Tanaka H. Early infection of scutellum tissue with *Agrobacterium* allows high-speed transformation of rice. The Plant Journal. 2006;47:969-976.

14. Li R, Qu R. High throughput *Agrobacterium*-mediated switch grass transformation. Biomass and bioenergy. 2010;30:1-9.

15. Sato H, Takamizo T. The protocol for Tall Fescue transformation. Pagg. 81-87, in [Transformation protocol in plants], Kagaku Doujin; 2012. (in Japanese).

3

Investigation of Genetic Diversity among *Medicago* species Using RAPD Markers

Komal Murtaza[1], Khushi Muhammad[1*], Mukhtar Alam[2], Ayaz Khan[1], Zainul Wahab[3], Muhammad Shahid Nadeem[1], Nazia Akbar[1], Waqar Ahmad[1] and Habib Ahmad[1]

[1]*Department of Genetics, Hazara University, Mansehra, Pakistan.*
[2]*Department of Plant Breeding and Genetics, University of Swabi, Swabi, Pakistan.*
[3]*Department of Art and Design, Hazara University, Mansehra, Pakistan.*

Authors' contributions

This work was carried out in collaboration between all authors. Authors KM and KM designed the study, performed the statistical analysis, wrote the protocol, and wrote the first draft of the manuscript. Authors MA, AK, and NA managed the analyses of the study. Authors MSN and WA managed the literature searches. Authors HA and ZA contributed in final manuscript writing. All authors read and approved the final manuscript.

<u>Editor(s):</u>
(1) Mahalingam Govindaraj, ICRISAT, Patancheru, India.
<u>Reviewers:</u>
(1) Zephaniah Dhlamini, Department of Applied Biology & Biochemistry, National University of Science & Technology, Zimbabwe.
(2) Ernestina Valadez-Moctezuma, Departamento de Fitotecnia. Universidad Autónoma Chapingo. México.
(3) Anonymous, China.

ABSTRACT

Aims: *Medicago* is known as the Queen of forage with potential economic importance to our society. The present study aimed at the use of RAPD-PCR DNA marker to identify the genetic fingerprints affinities of six species of Alfalfa.
Place and Duration of Study: The study was conducted at the Department of Genetics, Garden Campus, Hazara University, Mansehra Pakistan during February, 2011 to August, 2013.
Methodology: In this study, six species of *Medicago* namely *TWAL* (Tetraploid Wisconsin Alfalfa Line), *Medicago arborea, Medicago falcata, Medicago sativa, Medicago lupulina* and *Medicago polymorpha* were used to explore the diversity of alfalfa. Seven out of 120 decamers produced 34 polymorphic loci with 100% polymorphism to identify the different species of *Medicago* crop. The

---

Corresponding author: Email: khushisbs@yahoo.com

range of polymorphic loci was observed from 300 to 700 bp. Eleven species specific loci were generated by seven decamers. Primer B-18 generated single specific locus 700 bp against genomic DNA of *M. lupulina* and it is important to identify particular species of Alfalfa. The bivariate data were recorded as the presence of locus 1 and absence 0 and then this data was transferred into A and C respectively to make it suitable for DNAMAN software (version 5.2.2.0; Applied Biostatistics Inc). Moreover, cluster analysis was performed using sequence alignment and divergence function of the DNAMAN against the bivariate data collected from the products of decamers. All members clustered in a unique pattern except *M. falcata* and *M. lupulina* those shared 86% homology. Three distinct groups were observed during UPGMA (Unweighted pair Group Method with Arithmetic Mean). During the phylogenetic study, TWAL was observed to have genetic diversity from other five species of Alfalfa.

Conclusion: So, the present study is enabling us to discriminate different species of Alfalfa and it could be useful to identify and authenticate different species of the same genus of medicinal important plant from the Flora of Pakistan.

Keywords: Medicago; RAPD; Genetic diversity.

1. INTRODUCTION

Plants are essential constituents of ecosystems. Among plants, *Medicago* is one of the important legume plants that have been widely studied because it has numerous agriculturally important and domesticated species and known as queen of forage [1-2]. There are 87 species that have been recorded and these are distributed mainly around the Mediterranean basin [3-4]. A dozen annual species of *Medicago* have become significant pasture crops only in the past century, while best known species of this genus is *M. sativa* (alfalfa). It is the fourth most important crop in North America (after corn, soybean and wheat, respectively), and the temperate world's most important forage crop [5]. Several modern technologies are now being developed that have huge promise for extending the importance of alfalfa and its allied species well beyond their present primary use as a feed for livestock [5].

There are many effective and sensitive techniques which are being extensively used for characterization and conservation of crops and valuable plants [6-8]. Molecular marker can be employed for analysis of variance at the DNA level and different markers with useful genetic qualities (they can be dominant or co-dominant, can amplify anonymous or characterized loci, can contain expressed or non-expressed sequences, etc.) are being used by many researchers [9-12]. Analyses of genetic diversity are usually based on assessing the diversity of molecular markers, which tend to be selectively neutral [13].

Recently, many researchers have used the RAPD technique to estimate genetic diversity of various endangered plant species and valuable crops [14-18]. Random Amplified Polymorphic DNA (RAPD) is now in common use for ecological, evolutionary, taxonomy, phylogeny and genetic studies of plant sciences. These techniques are well established and their advantages and limitations have been documented properly [19].

In the present study, we attempted to discriminate, characterize and explore the genetics of valuable plants from Pakistan based on DNA markers and to increase the efficiency for study of crops at DNA level. It is the first comprehensive study at the DNA level of *Medicago* species from Pakistan using PCR based DNA markers. Our objective of the research was to identify specific regions of DNA linked to different species and establish clusters of species based on simple RAPD-PCR technique.

2. MATERIALS AND METHODS

2.1 Plant Material

Six species of *Medicago* namely *TWAL* (Tetraploid Wisconsin Alfalfa Line), *Medicago arborea*, *Medicago falcata*, *Medicago sativa*, *Medicago lupulina* and *Medicago polymorpha* were selected for the present study. Seeds of 3 species of *Medicago* were kindly provided by Prof. Habib Ahmad, Chairman, Department of Genetics, Hazara University, Mansehra. Hundred seeds of TWAL (Tetraploid Wisconsin Alfalfa Line), thirty-eight seeds of *Marborea* and hundred seeds of *M falcata* were planted in pots under greenhouse conditions at Garden Campus of Hazara University, Mansehra (Latitude

34°20'N, Longitude 7°15'E, Altitude 1066 meter) and leaves of 3 local varieties were also collected from the Garden Campus of Hazara University. For molecular study, one sample of each species was used to extract whole genomic DNA.

2.2 DNA Isolation

DNA was isolated from the leaves of *Medicago* species and two protocols were modified and optimized for DNA isolation from *Medicago,* Cetyl trimethyl ammonium bromide (CTAB) procedure [20] and a small scale DNA isolation procedure [21]. The quality and the quantity of DNA were estimated by capering with Fementas DNA molecular weight marker (cat. SM0403). The DNA was diluted to different concentrations (10 ng, 30 ng, 50 ng and 100 ng) in sterile double-distilled water for RAPD-PCR optimization reactions.

2.3 RAPD-PCR Primers

In the present study, RAPD primers were used against the genomic DNA of *Medicago* species to find genetic diversity. The primers used to identify the DNA marker linked to *Medicago* species were obtained from BIONEER.

2.4 RAPD PCR Optimization and Amplification of DNA Fragment

PCR mixture was prepared in sterile 200 µl tubes using commercially available kits (Wizbiosolutions, Korea), each mixture containing 50 ng of genomic DNA, 1x PCR buffer (75 mM Tris HCl PH 8.8 at 25°C, 20 mM $(NH4)_2$ SO_4 and 0.01 (Tween-200), 50 mM KCl, 2.5 mM $MgCl_2$, 200 µM deoxy nucleotide triphosphate (dNTPs; dATP, dCTP, dGTP, and dTTP, 10 pmol decamer arbitrary primers. The total volume of each reaction mixture was adjusted to 20 µl with sterilized water.

The amplification was carried out on DNA thermo cycler (Applied Biosystem., 2720 Thermocycler) with a modified version of the conditions used by William et al. [22]. It was programmed as single denaturation step for 5 min at 94°C followed by a step cycle programmed for 40 cycles of denaturation at 94°C for 1 min, annealing at 34 - 37°C (depending upon best amplification based on GC content of decamers) for 1 min and extension at 72°C for 2 min and final extension at 72°C for 10 min.

2.5 Resolving of PCR Product

All PCR products were resolved by electrophoresis along with the DNA marker on 1.5% agarose gels, prepared in 1X TAE (Tris/Acetate/EDTA) buffer. The electrophoretic file images were documented and saved using Uvitech gel documentation (gel documentation system).

2.6 Scoring and Processing of RAPD Data

The amplification of PCR products was done twice or thrice for reproducibility of band scoring. The size of amplified RAPD-DNA fragments (bp) was estimated by reference to a known DNA marker. Only reproducible PCR products were scored and used in subsequent analyses. The banding pattern of RAPD-PCR was scored for the presence (1) or absence (0) of specific amplicon and then 1 transferred into A and 0 into C as DNAMAN software can read data in the sequence form (Table 1). Data were analyzed using multiple sequence alignment and divergence function of DNAMAN statistical software (version 5.2.2.0; Applied Biostatistics Inc). Based on the divergence matrix and using the neighbor-joining method (NJ), we constructed a dendogram showing the difference between six as determined with 7 primers from 22 binary characters in the amplicon presence – absence [23]. The Nei's [24] genetic distance based on Unweighted pair Group Method with Arithmetic Mean (UPGMA).

3. RESULTS AND DISCUSSION

Medicago is considered as one the most important genera of forage plants [25] and also used as medicine, human food, green manure sources of industrial enzymes in biotechnology [26], model genomic species [27], and model systems for the study of nitrogen fixation [28], Six species of *Medicago* (*Fabaceae*) were used for DNA analysis. Total genomic DNA from six genetic stocks of polyploid *Medicago* was isolated using small scale DNA isolation procedure [20-21,29].

3.1 Amplification of RAPD Primers

The highly purified genomic DNA samples from six selected species were subjected to analysis and characterization of genomic synteny among them with the help of RAPD-PCR. One hundred and twenty RAPD markers were obtained from BioNeer kits and applied against the six DNA samples. After initial screening, 7 RAPD primers were chosen out of 120 primers for further study (Table 2). The selected 7 RAPD primers generated 34 scorable amplification products against six genomic DNA samples of *Medicago*. The results were analyzed by using DNAMAN software to generate homology and phylogenetic tree and various other parameters, i.e., total bands (TB), polymorphic bands (PB), monomorphic bands (MB), percentage of polymorphic (PP), were also considered.

3.2 Polymorphism and Polymorphic Content of Decamers

Polymorphism is the most important property of DNA marker, which can be used to discriminate the genetic basis of traits in crops. The seven selected primers produced 34 detectable amplicons in our DNA samples with the mean of 4.8 loci per primer. The number of amplified loci ranged from 1 to 9, with the approximate size range of 300 to 700bp and all used markers showed 100% polymorphism (Table 2). The executed RAPD markers showed 34 (100%) polymorphic loci in all six genotypes this property of the markers was potentially utilized to characterize different species of the family and genotypes of crops. Similar function of RAPD was reported in other valuable crops [30-31]. Observed polymorphic loci could be used as a tool to evaluate genetic diversity of species on the basis of presence and absence of specific locus [32,33], [18]. The least numbers of bands/loci i.e., one was amplified with primer B-18 (CCACAGCAGT), while the maximum number of loci i.e., mine was amplified with

primer C-19 (GTTGCCAGCC). Using polymorphic RAPD and ISSR markers, the variability analysis was recorded among the species of *Medicago* [34,35].

During the present study, three separate PCRs were carried out using one arbitrary primer B-18 in first reaction, three primers A-16, C-08 and C-18 in second and three primers C-19, D-10 and E-12 in the third reaction respectively (Table 3).

A similar approach was used and established proper identification system of genotypes and species of *Medicago* and various important plants in the study [36-38], [35]. The characterization of plant species is the most important objective of this study. Many researchers have reported genotype or variety specific loci generated by DNA markers identify verities of Potato [39], *Rhus* species [40] fig varieties [41], *Jatropha* genotypes [42], tea genotypes [43] and *Zingiber officinales* varieties [36].

3.3 Cluster Analysis

The UPGMA method classified six *Medicago* genotypes into different clusters having 61% to 86% homology (Fig. 1). In homology tree, *M. falcata* and *M. lupulina* clustered together having 86% homology while *M. arborea* shared 74% homology with *M. falcata* and *M. lupulina*. *M. polymorpha*, *M. sativa* and TWAL showed different branches with sharing 71%, 63%, 61% respectively homology to other three species (Fig. 1).

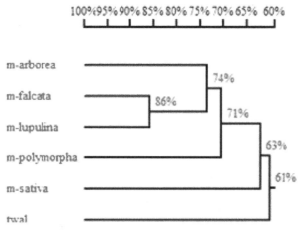

Fig. 1. Homology tree constructed against six accessions of *Medicago* species to determine genetic diversity. To draw a tree, to draw dendogram based on UPGMA, the DNAMAN 5.2.2.0 software was used. In the tree, TWAL, *M. sativa*, *M. polmorphia* and *M. arborea* showed separate branching while *M. falcata* and *M. lupulina* clustered together

Table 1. Bivariate (1-0) data matrix of 6 *Meicago* species using RAPD Primers A-16, B-18, C-08, C-18, C-19, D-10 and E-12

Alleles	TWAL	*M. arborea*	*M. falcate*	*M. sativa*	*M. lupulina*	*M. polymorpha*
1	0	1	0	0	0	0
2	0	0	0	1	0	1
3	0	0	0	1	0	0
4	0	0	0	0	1	0
5	0	0	0	0	0	1
6	0	0	0	1	0	1
7	1	1	0	0	0	0
8	0	0	0	1	0	0
9	0	0	0	0	0	0
10	0	0	1	1	0	1
11	0	1	0	0	0	0
12	1	0	0	0	0	0
13	0	0	1	0	0	0
14	1	1	1	1	1	0
15	1	0	0	1	0	0
16	0	1	0	0	0	0
17	1	0	0	0	0	0
18	1	0	0	0	0	0
19	0	0	0	1	0	0
20	1	1	0	1	0	1
21	1	0	0	0	0	0
22	0	0	0	1	0	0

Table 2. Details of decamer primers used in the present study

Primer name	Sequence (5'-3')	TB	MB	PB	PP%	Range of bands
A-16	AGCCAGCGAA	4	0	4	100%	400-500
B-18	CCACAGCAGT	1	0	1	100%	700
C-08	TGGACCGGTG	7	0	7	100%	350-700
C-18	TGAGTGGGTG	5	0	5	100%	350-700
C-19	GTTGCCAGCC	9	0	9	100%	300-550
D-10	GGTCTACACC	6	0	6	100%	300-600
E-12	TTATCGCCCC	2	0	2	100%	500-600
Total Bands		34	0	34		

Abbreviations: T.B = Total Bands, M.B = Monomorphic Bands, P.B = Polymorphic Bands, PP = Percentage of polymorphism

Table 3. Details of specific loci against different species of *Medicago* generated by selected RAPD primers

Primer	Characteristic band (bp)	Species					
		Arborea	Falcata	Lupulina	Polymorpha	Sativa	TWAL
A16	400	1	0	1	1	0	0
	485	0	0	1	0	0	0
B18	550	0	1	0	0	1	1
C-08	300	0	0	0	0	0	1
	400	1	0	0	0	0	0
	500	0	0	0	0	1	0
	700	0	0	0	1	0	0
C-18	300	0	0	0	0	0	1
	600	1	0	0	0	0	0
	700	0	1	0	0	0	0
Total	11						

Using PCR based DNA markers, genetic diversity was investigated among the species or varieties of alfalfa and different valuable crops [34,35,44,45]. The genetic affinities were investigated based on data generated by RAPD marker and this approach was widely used during previous research [46]. This study could be useful to evaluate taxonomic positions of members of *Medicago* and it is providing a baseline for future to develop different molecular techniques to characterize different species from Pakistan.

4. CONCLUSION

In the present, the polymorphic properties of RAPD marker were utilized and 11 species specific loci were identified. To evaluate genetic diversity of *Medicago* species, cluster analysis was performed and 3 clusters were identified and genetic divergence ranging 61% to 86% among six species was recorded. Here, our study is concluding that *M. falcata* and *M. lupulina*. *M. polymorpha*, *M. sativa* and TWAL had a unique pattern of clustering. This study could be able to provide information based on DNA fingerprints.

ACKNOWLEDGEMENTS

We are grateful to the Higher Education Commission of Pakistan for providing funding to complete this research and Prof. Javed Iqbal, School of Biological Sciences, University of the Punjab, Lahore, Pakistan for his kind assistance.

COMPETING INTERESTS

Authors have declared that no competing interests exist.

REFERENCES

1. Small E. Alfalfa and relatives: Evolution and classification of Medicago.NRC Research Press, Ottawa, Ontario, Canada. 2010.

2. Lacefield GD, Ball DM, Hancock D, Andrae J, Smith R. Growing Alfalfa in the South. NAFA. 2009;1-14.

3. Steele KP, Ickert-Bond SM, Zarre S, Wojciechowski MF. Phylogeny and character evolution in *Medicago* (*Leguminosae*): Evidence from analyses of plastid *trnK/ matK* and nuclear *GA3ox1* sequences. Am J Bot. 2010;97(7):1142–1155.

4. Gholami A, De Geyter N, Pollier J, Goormachtig S, Goossens A. Natural product biosynthesis in *Medicago* species. Natural Product Reports. 2014;31(3):356–380. DOI:10.1039/C3NP70104B.PMID2448147 7.

5. Small E. Alfalfa and relatives: Evolution and classification of *Medicago*. NRC Research Press, Ottawa, Ontario, Canada. 2011;pp 727.

6. Shengji P. Ethno botanical approaches of traditional medicine studies: some experiences from Asia. Pharm Bot. 2001; 39:74-79.

7. Samin J, Khan MA, Siraj-ud-din, Murad W, Hussain M, Ghani A. Herbal remedies used for gastrointestinal disorders in Kaghan Valley, NWFP, Pakistan. Pak J Weed Sci Res. 2008;14:169-200.

8. Pant S, Samant SS. Diversity, distribution, uses and conservation status of plant species of the Mornaula Reserve Forests, West Himalaya, India. Int J Biodivers Sci Management. 2009;2:97–104.

9. Xue CY, Li DZ. Molecular identification of the traditional Tibetan medicinal plant *Gentianopsis paludosa* (Gentianaceae) using diagnostic PCR and PCR-RFLP based on nrDNA ITS regions. Planta Med. 2007;73:610. DOI: 10.1055/s-2007-987390.

10. Mondini L, Arshiya N, Pagnotta MA. Assessing plant genetic diversity by molecular tools. Department of Agrobiology and Agrochemistry, Tuscia University. 2009;1:19-35.

11. Kool A, Boer HJD, Kruger A, Rydberg A, Abbad A. Molecular identification of commercialized medicinal plants in Southern Morocco. PLoS ONE. 2012;7(6): e39459. DOI:10.1371/journal.pone.0039459.

12. Saidi M, Movahedi K, Mehrabi AA. Characterization of genetic diversity in *Satureja bachtiarica* germplasm in Ilam province (Iran) using ISSR and RAPD markers. Intl J Agri Crop Sci. 2013;5(17):1934-1940. IJACS/2013/5-17/1934-1940.

13. Reed DH, Frankham R. Correlation between fitness and genetic diversity. Cons Biol. 2003;17:230-237.

14. Wang ZS, An SQ, Liu H, Leng X. Genetic structure of the endangered plant *Neolitsea sericea* (*Lauraceae*) from the

DOI:10.3732/ajb.1000009.PMID 21616866

Zhoushan archipelago using RAPD markers. Ann Bot. 2005;95:305-313.

15. Lu HP, Cai YW, Chen XY, Zhang X. High RAPD but no cpDNA sequence variation in the endemic and endangered plant, *Heptacodium miconioides* Rehd. (*Caprifoliaceae*). Genetica. 2006;128:409-417.

16. Liu P, Yang YS, HaoC Y, Guo WD. Ecological risk assessment using RAPD and distribution pattern of a rare and endangered species. Chemosphere. 2007;68:1497-1505.

17. Zheng W, Wang L, Meng L, Liu J. Genetic variation in the endangered *Anisodus tanguticus* (*Solanaceae*), an alpine perennial endemic to the Qinghai-Tibetan Plateau. Genetica. 2008;132:123-129.

18. Muhammad K, Afghan S, Pan YB, Iqbal J. Genetic variability among the brown rust resistant and susceptible genotypes of sugarcane by rapid technique. Pak J Bot. 2013;45:163-168.

19. Agarwal M, Shrivastava N, Padh H. Advances in molecular marker techniques and their application in plant sciences. Plant Cell Rep. 2008;27:617-631.

20. Doyle JJ, Doyle JL. Isolation of plant DNA from fresh tissue. Focus. 1990;12:13-15.

21. Weining S, Langridge P. Identification and mapping of polymorphism in cereal base on polymerase chain reaction. Theor Appl Genet.1992;82:209-216.

22. Williams JGK, Kubelik AR, Livak KJ, Rafalski JA, Tingey SV. DNA polymorphisms amplified by arbitrary primers are useful as genetic markers. Nucleic Acids Res. 1990;18:6531-6535.

23. Saitou N, Nei M. The neighbor-joining method: a new method for reconstructing phylogenetic trees. Mol Biol Evol. 1987;4:406-425.

24. Nei M. Estimation of average heterozygosity and genetic distance from a small number of individuals. Genetics. 1978;89:583-590.

25. Heyn CC. The Annual Species of *Medicago*. *Scripta hierosolymitana*, Hebrew University Press, Jerusalem; 1963.

26. Lewis G, Schrire B, Mackinder B, Lock M. Legumes of the world. Royal Botanic Gardens, Kew, UK; 2005.

27. Cannon SB, Sterck L, Rombauts S, Sato S, Cheung F, Gouzy J, Wang X. Legume genome evolution viewed through the *Medicago truncatula* and Lotus Japonicus genomes. P Natl Acad Sci USA. USA. 2006;103:14959–14964.

28. Bailly X, Olivieril, Brunel B, Cleyet-Marel JC, Bena G. Horizontal gene transfer and homologous recombination drive the evolution of the nitrogen-fixing symbionts of *Medicago* species. J. Bacteriol. 2007; 189:5223–5236.

29. Lodhi MA, Guang NY, Norman FW, Bruce IR. A simple and efficient method for DNA extraction from grapevine cultivars and Vitis species. Plant Mol Biol Rep. 1994; 12:6–13.

30. Khan FA, Khan A, Azhar FM, Rauf S. Genetic diversity of *Saccharum officinarum* accessions in Pakistan as revealed by random amplified polymorphic DNA. Genet Mol Res. 2009;8(4):1376-1382.

31. Ali W, Muhammad K, Nadeem MS, Inamullah, Ahmad H, Iqbal J. Use of RAPD markers to characterize commercially grown rust resistant cultivars of sugarcane. Int J Biosci. 2013;3(2):115-12.

32. Solouki M, Mehdikhani H, Zeinali H, Emamjomeh AA. Study of genetic diversity in chamomile (*Matricaria chamomilla*) based on morphological traits and molecular markers. Sci Hortic. 2008; 117(3):281-7.

33. Okon S, Surmacz-Magdziak A. The use of RAPD markers for detecting genetic similarity and molecular identification of chamomile (*Chamomilla recutita* (L.) Rausch.) Genotypes. Herbica Polonica. 2011;57(1):39-47.

34. Brummer EC, Bouton JH, Kochert G. Analysis of annual Medicago species using RAPD markers. Genome. 1995;38(2):362-7.

35. Xavier JR, Kumar J, Srivastava RB. Characterization of genetic structure of alfalfa (*Medicago* sp.) from trans-Himalaya using RAPD and ISSR markers. Afr J Biotechnol. 2011;10(42):8176-8187.

36. Pal MD, Raychaudhuri SS. Estimation of genetic variability in *Plantago ovata* cultivars. Biol Plantarum. 2003;47:459-462.

37. Bauvet JM, Fontaine C, Sanou H, Cardi C. An analysis of the pattern of genetic variation in *Vitellaria paradoxa* using RAPD marker. Agroforest Syst. 2004;60:61-69.

38. Harisaranraj R, Suresh K, Saravanababu S. DNA finger printing analysis among eight varieties of *Zingiber officinale* rosc. by using RAPD markers. Global J Mol Sci. 2009;4:103-107.

39. Prevost A, Wilkinson MJ. A new system of comparing PCR primers applied to ISSR fingerprinting of potato cultivars. Theor Appl Genet. 1999;98(1):107-12.

40. Prakash S, Van SJ Assessment of genetic relationships between Rhus L. species using RAPD markers. Genet Res Crop Evol. 2007;54(1):7-11.

41. Baraket G, Chatti K, Saddoud O, Mars M, Marrakchi M, Trif IM, Salhi HA. Genetic analysis of Tunisian fig (*Ficus carica* L.) cultivars using amplified fragment length polymorphism (AFLP) markers. Sci Hortic. 2009;120:487-92.

42. Tatikonda L, Wani SP, Kannan S, Beerelli N, Sreedevi TK, Hoisington DA, Devi P, Varshney RK. AFLP-based characterization of an elite germplasm collection of *Jatropha curcas* L., a biofuel plant. Plant Sci. 2009;176(4):505-13.

43. Chen L, Gao Q, Chen D, Xu C. The use of RAPD markers for detecting genetic diversity, relationship and molecular identification of Chinese elite tea genetic resources [*Camellia sinensis* (L.) O. Kuntze] preserved in a tea germplasm repository. Biodivers Conserv. 2009;14(6): 1433-44.

44. Saengprajak J, Saensouk P. Genetic diversity and species identification of cultivar species in *Subtribe cucumerinae* (Cucurbitaceae) using RAPD and SCAR markers. Am J Plant Sci. 2012;3:1092-1097.

45. Surgun Y, Col B, Burun B. Genetic diversity and identification of some Turkish cotton genotypes (*Gossypium hirsutum* L.) by RAPD-PCR analysis. Turk J Biol. 2012;36:143-150.

46. Mahmood Z, Raheel F, Dasti AA, et al. Genetic diversity analysis of the species of Gossypium by using RAPD markers. Afr J Biotechnol. 2009;8:3691-3697.

4

Molecular Analysis of Genetic Diversity of *Pectobacterium carotovorum* Subsp *Carotovorum* Isolated in Morocco by PCR Amplification of the 16S-23S Intergenic Spacer Region

M. Amdan[1], H. Faquihi[1*], M. Terta[1], M. M. Ennaji[1] and R. Ait Mhand[1]

[1]*Laboratory of Virology, Microbiology and Quality / Ecotoxicology and Biodiversity, University Hassan II Mohammedia-FSTM, Morocco.*

Authors' contributions

This work was carried out in collaboration between all authors. Author MA performed experimental work and wrote the first draft of the manuscript. Author HF revised and completed the manuscript, author MT performed the data analysis, author MME contributed to the supervision of this work and author RAM designed, supervised, managed the work and revised the final manuscript. All authors read and approved the final manuscript.

Editor(s):
(1) Giuliana Napolitano, Department of Biology (DB), University of Naples Federico II, Naples, Italy.
Reviewers:
(1) Amira Souii, Department of Microbiology and Biotechnology, University of Tunis El Manar, Tunisia.
(2) Ilham Zahir, Department of Biology, Faculty of Sciences and Technology, University Sidi Mohamed Ben Abdellah, Morocco.
(3) Vijai Singh, Institute of Systems & Synthetic Biology, Université d'Évry Val d'Essonne, France.

ABSTRACT

Aims: *Pectobacterium carotovorum* is a ubiquitous bacterium that causes soft rot in different crop plants throughout the world. In Morocco, approximately 95% of the Strains isolates from potato plants with tuber soft rot are *P. carotovorum*. In this study, we test whether PCR ribotyping can be used to distinguish strains of *Pectobacterium carotovorum* isolated from soft rot potato and to differentiate among strains from different geographic regions.
Place and Duration of Study: Laboratory of Virology, Microbiology and Quality / Ecotoxicology and Biodiversity, department Biology, Faculty of Sciences and Techniques, University Hassan-II Mohammedia Casablanca.
Methodology: Eighty-three pectolytic enterobacteria were collected from potatoes rotten in Morocco, the strains were isolated in the Cristal Violet Pectate (CVP) medium and were purified in

LPGA agar (yeast extract, peptone, glucose and agar). After purification, strains were identified by physiological and biochemical tests. The confirmation of species was performed by PCR using primers Y1 and Y2. The genetic diversity of *Pectobacterium carotovoum* was investigated by PCR ribotyping using primers G1/L1, which are complementary to conserved regions of the rRNA operon. Furthermore, the profiles obtained were compared by the Unweighted Pair Group Method.
Results: The biochemical and physiological analysis demonstrated that the predominant pectolytic enterobacterium present in Morocco is *Pectobacterium carotovorum* subsp *carotovorum*. The specific confirmation of species *P. carotovorum* by PCR has yielded a 434 bp DNA fragment of the pelY gene with all isolates. Further, PCR amplification of the 16S-23S Intergenic spacer Region (ITS-PCR) has presented a specific pattern made of 2-6 fragments ranging from 300 bp to 800 bp. The UPGMA tree has shown that there is considerable genetic diversity in *P. carotovorum* strains, which can be divided into four distinct groups.

Keywords: *Pectobacterium carotovorum; potato soft rot; internal transcribed spacer (ITS); genetic diversity.*

1. INTRODUCTION

Soft rot bacteria are global pathogens and amongst the most prevalent and destructive bacterial diseases that affect potato, particularly during storage and transport [1]. It is mainly due to the phytopathogenic agent Pectobacterium, which belongs to the *Enterobacteriaceae* family. This genus has been subjected to extensive taxonomic research and consequently, divided into several species and subspecies on the basis of molecular and biochemical differences [2-4]. To date, three Pectobacterium species have been described: Pectobacterium atrosepticum (Pa), Pectobacterium was abiae (Pw) and *Pectobacterium carotovorum* (Pc) [4,5]. The latter is presented by two subspecies were described as organisms causing potato soft rot. *Pectobacterium carotovorum* sub sp. *carotovorum,* responsible for 95% of soft rot disease in Morocco [6] and *Pectobacterium carotovorum* sub sp. brasiliensis, a highly aggressive bacterium, responsible for the majority of blackleg incidences in Brazil [7], in South Africa [8] and was lately discovered in Morocco [9]. *Pectobacterium wasabiae* has been described as a new potato pathogen in Canada [10] and was responsible for the high blackleg levels in New Zealand [11] and also in Japan [4,11,12]. As a result, the lot of species and sub-species associated with soft rot disease makes identification of the causal agent by phenotypic symptoms nearly impossible. Lately, several typing methods based on molecular approaches have been developed for the study of *Pectobacterium* ssp, such as PCR amplification and sequencing, RFLP (restriction fragment length polymorphism) of the 16S gene and the 16S-23S rDNA intergenic spacer [13,14], RFLP

of recA gene fragments [15,16] and AFLP fingerprinting [17]. In prokaryotes, the conserved macromolecules 16S rRNA has been widely used for bacterial identification; classification and phylogenetic analysis [18]. This is because the 16S rRNA gene has important length ($\approx$ 1500pb) and is present in all bacteria [19]. The rRNA locus rrn is highly conserved and consequently can be employed to characterize the evolutionary correlation. The rrn loci contain the genes for all three rRNA subunits, 16S, 23S and 5S genes and they are separated by a large number of various sequences and intergenic spacer regions [20]. ITS is adapted to differentiate strains within a certain species and used as a marker of the subspecies [21,22]. The different species of prokaryotes can be distinguished by the polymorphisms of sequences and length of ITS [23].

In order to evaluate the efficiency of ITS method to identify *P. carotovorum*, we used the PCR ribotyping for the purpose of developing a rapid, accurate identification and a phylogenetic analysis of this pathogen. Thus, the aims of this work were (1) to detect, identify and characterize *P. carotovorum* isolated from soft rot potato by physiological, biochemical and molecular methods; (2) to use ITS-PCR analysis for identification of strains and to study intra-species diversity; (3) to determine whether ITS-PCR fingerprinting could be an useful tool for the characterization of the genetic variability among isolates of *P. carotovorum*. Thereby, the study was conducted with a survey of potato production areas and suspected samples were collected, analyzed and characterized by both biochemical and molecular methods. Then, PCR amplification of the 16S-23S intergenic spacer region of these strains was carried out and revealed a considerable genetic variability of *P. carotovorum*.

2. MATERIALS AND METHODS

2.1 Isolation of Bacterial Strains

Eighty-three *Pectobacterium* sp. strains were isolated from a range of potato cultivars in Morocco. In brief, bacterial cultures were isolated from the margin of infected tuber samples and spread as serial dilutions onto CVP medium at 27°C for 48–72 h. Those bacteria that caused cavities within 48 h were transferred onto Kings B medium and incubated at 27°C for 24 h. Collection of 83 strains used in this study had been stored after purification in 20% glycerol at-80°C [6,24]. In this experiments, the strains were grown in LPGA agar (5 g/L yeast extract, 5 g/L peptone, 5 g/L glucose 15 g/L agar) at 27°C during 16-18 h.

2.2 Biochemical and Phenotypic Tests

Suspected colonies typical Pectobacterium sp. were subjected to preliminary tests: Gram staining, catalase, oxidase, nitrate reduction and the test of rot on potato's slices. Thereafter, Identification of confirmed *Pectobacterium* sp. isolates to species and subspecies was conducted on the basis of the physiological and biochemical characteristics such as: Indole production from tryptophan, acid production from (lactose, trehalose, α-methyl-glucoside and melibiose), growth in 5% NaCl, growth on nutrient agar at 37°C, production of reducing substances from sucrose and lecithinase activity [25,26]. All tests were carried out at 27°C for 24 h and compared with the standard strains (reference strain of *Pectobactrerium carotovorum* 132C) [27].

2.3 Preparation of Bacterial DNA

DNA extraction was performed on 700 µL of the bacterial suspension 108 CFU. mL-1 of a pre-culture of 24 h, according to method of Li and De Boer [28]. The precipitated DNA was quantified, adjusted to 100 ng/µl using a Nano Drop 8000 spectrophotometer (Thermo Scientific, Wilmington, DE, USA) and stored at 4°C.

2.4 Molecular Detection by Primers Y1/Y2

Primers Y1/Y2 were used to amplify a 434 bp fragment of a pectate lyase encoding gene (pel gene) as described previously to identify *Pectobacterium carotovorum* by Darrasse et al. [29]: Y1 (5'-TTA CCG GAC GCC GAG CTG TGG CGT-3') and Y2 (5'-CAG, GAA, GAT, GTC, GTT, ATC, GCG, AGT-3'). The PCR was performed in 25 µl of a reaction mixture containing 0.4 µM of each primer, 2.5 µl of 10×PCR buffer, 1.5 mm MgCl2, 200 µM of dNTP, 1U of Taq polymerase (Promega) and 100 ng of template DNA. PCR conditions were performed using thermal cycler (Perkln Elmer Cltus) as follows: Initial denaturation step 95°C for 1 min, 35 cycles of denaturation 94°C for 1 min, annealing temperature 65°C for 1 min and extension at 72°C for 1.5 min, followed by a final extension step at 72°C for 10 min. PCR products were subjected to electrophoresis in 1.5% agarose gel in TBE buffer. Following staining with ethidium bromide, the gels were viewed and photographed under UV transilluminator. A standard 100 pb DNA ladder (Promega) was included on each gel as well as positive and negative controls.

2.5 ITS-PCR Amplification

The 16S-23S intergenic transcribed spacer was amplified using the PCR conditions and primers G1 (5'GAAGTCGTAACAAGG-3') and L1 (5'CAAGGCATCCACCGT-3') as described previously by Toth et al. [12]. Briefly, 25 µl of reaction mixture contained; 2.5 µl of 10×PCR buffer, 2 mm MgCl2, 200 µM of deoxynucleoside triphosphates, 0.4 µM of each primer (G1 and L1) and 1U of Taq polymerase and 1 µl of template DNA. The amplification conditions were as follows: Initial denaturation step at 94°C for 5 min, 28 cycles of denaturation 94°C for 1 min, annealing temperature 55°C for 2 min, extension at 72°C for 2 min, followed by a final extension step at 72°C for 7 min. The amplicons of ITS-PCR were separated by electrophoresis on 1.2% agarose gel containing 0.5 µg/ml ethidium bromide. The gels were visualized and photographed under UV transilluminator. Fragment sizes were determined by comparison to a DNA ladder.

2.6 Data Analysis

ITS-PCR fingerprints of amplified DNA fragments obtained from the agarose gel electrophoresis were recorded. The observed bands in the gels were evaluated based on the presence (coded 1) or absence (coded 0) of polymorphic fragments. Cluster analysis was performed with UPGMA using the online dendrogram construction utility, Dendro UPGMA(http://genomes.urv.cat/UPGMA) [30].

3. RESULTS AND DISCUSSION

Bacterial soft rot is one of the most devastating diseases of Morocco potatoes. The disease is characterized by foul smelling rot and collapse of the potato tubers. The bacterium Pectobacterium proves a very important role in disease incidence in Morocco. However, the emergence of several sub-species associated with soft rot disease makes identification of causal agent by looking at the symptoms nearly impossible. Hence, the biochemical and genetic methods of identification are required in order to accurately identify the relevant causal agent. Concerning this, the goal of this study was to develop and optimize a method for the simple, rapid and accurate identification of *P. carotovorum*. This was accomplished by using a combination of ITS-PCR, as well as morphological characteristics, biochemical and phenotypic tests. A collection of eighty three isolates has induced round cavities, 2-3 mm in diameter and has presented whitish grey coloration on CVP medium indicating the degradation of polygalacturonic acid. In addition, all the strains did not present pigmentation in the King B medium (Difco) and they were able to produce soft rot on potato slices. The biochemical properties of 83 enterobacteria infecting potatoes in Morocco have revealed that all are *P. carotovorum* sub sp. carotovorum. Furthermore, all genomic DNA were successfully amplified by inducing an approximately 434 pb fragment corresponding to pectate lyase genes belonging to the Y family of *P. carotovorum* (Fig. 1).

PCR amplification of the 16S-23S rRNA intergenic transcribed spacer region (ITS) using L1 and G1 primers produced 13 different DNA patterns in the 83 analyzed strains. All the profiles were characterized by the presence of bands whose size varied between 300 to 800 bp (Fig. 2), Furthermore, the bands 530,580,600 and 620bp were frequently observed in the majority of the profiles (Fig. 2). This finding suggested that there is considerable genetic diversity in *P. carotovorum* strains, which is in accordance with previous works reported by Eisa et al. [31].

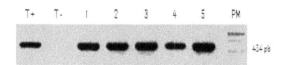

Fig. 1. Agraose gel of PCR-amplification of *Pectobacterium carotovorum* DNA by primers Y1 and Y2
The product was DNA fragments of 434pb PM: DNA ladder; T+:positive control; T-:negative control and lanes 1 to 5 isolates of the collection

On the other hand, The ITS-PCR gel picture was converted from JPEG to TIFF format and was subjected to (UPGMA) analysis. A dendrogram was then constructed (Fig. 3) and the relationship between isolates was estimated. The eighty-three profiles obtained were regrouped in 19 elements with 18 variables and each element has been analyzed. The UPGMA analysis was categorized the strains in four clusters with tree minor clusters and one majority: A; comprising all isolates and the three other clusters B, C and D including each one strain. Moreover, ten sub-clusters could be identified within cluster A: eight groups showed the same pattern, clustering at 100% similarity (The strains 10, 11, 14, 15, 18, 19, 20, 25, 31, 32, 37, 49, 50, 53, 54, 57, 58, 59, 61, 62, 64, 70, 73, 74, 75, 77 were identical to 6, The strains 3, 7, 9, 12, 21, 24, 30, 36, 45, 46, 51, 63, 65, 66, 67, 69, 72 were identical to 2, The strains 55,56,60 were identical to 48, The strains 23, 28, 29, 33, 76, 78 were identical to 8, The strains 41,42,44 were identical to 38, The strain 5 was identical to 4, The strains 68, 71 were identical to 1 and The strain 22 was identical to 13) and the rest of strains showed own and distinct patterns.

Because the symptoms caused by *P. carotovorum* resemble those of soft rot caused by other soft rot bacteria, The development of a specific, rapid diagnostic method for soft rot bacteria is noteworthy with regard to import and export regulations for farm products. Results using primers Y1/Y2, which amplified the expected bands from the isolates, were in accord with the classification based on physiological and biochemical features, and *P. carotovorum* was generally isolated from the infected potato. Thereby, these data reflected that *P. carotovorum* is the major bacterium responsible of soft rot disease in Moroccan potatoes. These findings strongly correlated with the results found in others studies in Morroco [23], in Bangladeshi [32], in Iran [33], in Malaysia [34] and in korea

[35]. This species is prevalent on potatoes worldwide [36-38]. The pervasiveness of *P. carotovorum* may be due to an earlier divergence, wider geographical distribution or broader host range and is thought to have resulted in the extensive genetic variability of this subspecies [16].

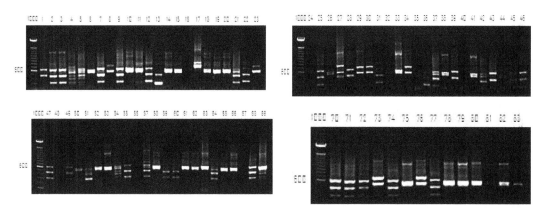

Fig. 2. Genomic DNA fingerprinting patterns from strains of *Pectobacterium carotovorum* isolated from potato, generated by ITS-PCR
1000: marker (100bp); lanes 1 to 83: strains of Pectobacterium carotovorum

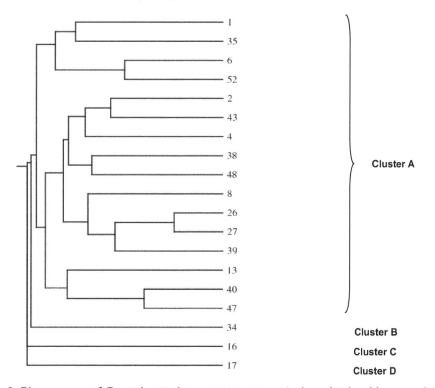

Fig. 3. Phenogram of *Pectobacterium carotovorum* strains obtained by unweighted pair Group method based on the whole ITS-PCR data set. Average arithmetic clustering (UPGMA) and visual examination were used to identify groups. A, B, C and D represent the clusters obtained
The strains 10,11,14,15,18,19,20,25,31,32,37,49,50,53,54,57,58,59,61,62,64,70,73,74,75,77 were identical to 6; the strains 3,7,9,12,21,24,30,36,45,46,51,63,65,66,67,69,72 were identical to 2; the strains 55,56, 60 were identical to 48; the strains 23,28, 29,33,76,78 were identical to 8; the strains 41, 42, 44 were identical to 38; the strain 5 was identical to 4; the strains 68 and 71 were identical to 1; the strain 22 was identical to 13

In addition, to obtain more precise identification of isolates, ITS-PCR profiles were generated using primers complementary to the 3' end of the 16S and 5' end of the 23S rRNA genes resulted in DNA fragments between 300-800 bp. Hence, the multiple banding patterns produced with ITS-PCR from different isolates was also observed by Toth et al. [12]. Moreover, sizes and sequences of the large and small ITS regions of Moroccan isolates were also similar to those of other *P. carotovorum* found in other work where the ITS-PCR fragments obtained were from 300 to 600 bp [39].

Noting that, two to 11 copies of the rRNA loci are present on the chromosomes of most bacterial species and, in some bacterial groups, the different copies of the intergenic spacer regions show extensive length variation [40,41]. A minimum of two to four copies of the rrn locus have been described in P. atrosepticum and P. betavasculorum, all of them having nearly identical length and sequence [12]. In our hands, two to six amplified ITS fragment was obtained for each strain, confirming there is high genetic variation in the ITS region within each strain of *P. carotovorum*. Indeed, intergenic rDNA sequences are known to have higher variability than functional sequences of the rrn cistrons, allowing the distinction of bacteria below the species level in some cases [42]. Also, these sequences have several advantages, including being highly conserved in *P. carotovorum* and easy to amplify. On this basis, a higher degree of differentiation was expected from PCR of the 16S–23S rDNA intergenic spacer, as have been reported by other studies [12,24,43-45] using several molecular analyses seeking to understand the relationship among the Pectobacterium sp. strains.

Finally, our study has identified a possible new tool for detection of levels of diversity amongst pectolytic enterobacteria *P. carotovorum* causing soft rot in morocco, which may have important implications for their detection elsewhere.

4. CONCLUSION

An objective of this study was to evaluate the current methods for detection. The results indicated that the higher variability of the ITS sequence and length in *P. carotovorum* bacteria can be used to identify strains or related groups of strains, while it is not extensible to rapid species identification, because of the lack of pattern identity among strains belonging to the same species. Thereby, the analysis of the 16S-23S rRNA gene internal transcribed spacer (ITS) is a reliable method to classify strains of *Pectobacterium carotovorum* and able to differentiate between strains using their high diversity of polymorphism.

ACKNOWLEDGEMENTS

This work was supported by project ARIMNET PoHMed French and Moroccan cooperation.

COMPETING INTERESTS

Authors have declared that no competing interests exist.

REFERENCES

1. Czajkowski R, Grzegorz G, Van Der Wolf JM. Distribution of *Dickeya* spp. and *Pectobacterium carotovorum* subsp. *carotovorum* in naturally infected seed potatoes. Eur J Plant Pathol. 2009;125: 263-75.

2. Kwon SW, Go SJ, Kang HW, Ryu JC, Jo JK. Phylogenetic analysis of *Erwinia* species based on 16S rRNA gene sequences. Int J Syst Bacteriol. 1997;47: 1061–67.

3. Hauben L, Moore ER, Vauterin L, Steenackers M, Mergaert J, Swings J, et al. Phylogenetic position of phytopathogens within the *Enterobacteriaceae*. Syst Appl Microbiol. 1998;21:384–97.

4. Gardan L, Gouy C, Christen R, Samson R. Elevation of three subsp. of *Pectobacterium carotovorum* to species level: *Pectobacterium atrosepticum* sp. nov. *Pectobacterium betavasculorum* sp. nov. and *Pectobacterium wasabiae* sp. nov. Int J Syst Evol Microbiol. 2003;53: 381–91.

5. De Haan Eisse G, Dekker-Nooren Toos CEM, Van Den Bovenkamp GW, Speksnijder Arjen GCL, Van Der Zouwen PS, Van Der Wolf JM. *Pectobacterium carotovorum* subsp. *carotovorum* can cause potato blackleg in temperate climates. Eur J Plant Pathol. 2008;122: 561–69.

6. Terta M, El Karkouri A, Ait Mhand R, Achbani E, Barakate M, Ennaji MM, et al. Occurrence of *Pectobacterium carotovorum* strains isolated from potato soft rot in Morocco. Cell Mol Biol. 2010; 56:1324-33.

7. Duarte V, De Boer SH, Ward LJ, De Oliveira AMR. Characterization of atypical *Erwinia carotovora* strains causing blackleg of potato in Brazil. J Appl Microbiol. 2004;96:535–45.

8. Van Der Merwe JJ, Coutinho TA, Korsten L, Van Der Waals JE. *Pectobacterium carotovorum* subsp. *brasiliensis* causing blackleg on potatoes in South Africa. Eur J Plant Pathol. 2010;126:175–85.

9. Terta M. Etude de la Biodiversité Biochimique Et Moléculaire du *Pectobacterium* sp. Agents Responsables des Pourritures Molles de Pomme de Terre au Maroc. Ph. D. Thesis. Faculty of science and technology Mohammedia, Morroco; 2011.

10. De Boer SH, Li X, Ward LJ. *Pectobacterium* spp. associated with bacterial stem rot syndrome of potato in Canada. Phytopathol. 2012;102:937–47.

11. Pitman AR, Harrow SA, Visnovsky SB. Genetic characterisation of *Pectobacterium wasabiae* causing soft rot disease of potato in New Zealand. Eur J Plant Pathol. 2010;126:423–35.

12. Czajkowski R, Perombelon MCM, Van Veen JA, Van Der Wolf JM. Control of blackleg and tuber soft rot of potato caused by *Pectobacterium* and *Dickeya* species. Plant Pathol. 2011;60(6):999–1013.

13. Toth IK, Avrova AO, Hyman LJ. Rapid identification and differentiation of the soft rot *Erwinias* by 16S-23S intergenic transcribed spacer-PCR and restriction fragment length polymorphism analyses. Appl Environ Microbiol. 2001;67:4070-76.

14. Fessehaie A, De Boer SH, Levesque CA. Molecular characterization of DNA encoding 16S-23S rRNA intergenic spacer regions and 16S rRNA of pectolytic *Erwinia* species. Can J Microbiol. 2002;48:387-398.

15. Waleron M, Waleron K, Podhajska AJ, Ojkowska E. Genotyping of bacteria belonging to the former *Erwinia* genus by PCR-RFLP analysis of a recA gene fragment. Microbiol. 2002;148:583-95.

16. Bahnaz R, Nader H, Javad R, Abolghasem G. Genetic diversity of Iranian potato soft rot bacteria based on polymerase chain reaction-restriction fragment length polymorphism (PCR RFLP) analysis. Afr J Biotechnol. 2012;11(6):1314-20.

17. Avrova AO, Hyman LJ, Toth RL, Toth IK. Application of amplified fragment length polymorphism fingerprinting for taxonomy and identification of the soft rot bacteria *Erwinia carotovora* and *Erwinia chrysanthemi*. Appl Environ Microbiol. 2002;68:1499-508.

18. Mignard S, Flandrois JP. 16S rRNA sequencing in routine. Bacterial identification: A 30-month experiment. J Microbiol Methods. 2006;67(3):574-81.

19. Clarridge JEIII. Impact of 16S rRNA gene sequence analysis for identification of bacteria on clinical microbiology and infectious diseases. Clin Microbiol Rev. 2004;17:840-62.

20. Cedergren R, Gray MW, Abel Y, Sankoff D. The evolutionary relationships among known life forms. J Mol Evol. 1988;28:98-112.

21. Frothingham R, Wilson KH. Molecular phylogeny of the *Mycobacterium avium* complex demonstrates clinically meaningful divisions. J Infect Dis. 1994; 169:305–12.

22. Gürtler V, Stanisch VA. New approaches to typing and identification of bacteria using the 16S-23S rDNA spacer region. Microbiol. 1996;142:3–16.

23. Zhu L, Xie H, Chen S, Ma R. Rapid isolation, identification and phylogenetic analysis of *Pectobacterium carotovorum* ssp. J Plant Pathol. 2010;92(2):479-83.

24. Kettani-Halabi M, Terta M, Amdan M, El Fahime E, Bouteau F, Ennaji MM. An easy, simple inexpensive test for the specific detection of *Pectobacterium carotovorum* subsp. *carotovorum* based on sequence analysis of the pmrA gene. BMC Microbiol. 2013;13:176.

DOI: 10.1186/1471-2180-13-176.

25. Fassihiani A, Nedaeinia R. Characterization of Iranian *Pectobacterium carotovorum* strains from sugar beet by phenotypic tests and whole-cell proteins profile. J Phytopath. 2008;156:281–86.

26. Barabote RD, Johnson OL, Zetina E, San Francisco SK, Fralick JA, San Francisco MJ. *Erwinia chrysanthemi* tolC is Involved in Resistance to Antimicrobial Plant Chemicals and is Essential for Phytopathogenesis. J Bacteriol. 2003; 85(19):5772–8.

27. Anajjar B, Azelmat S, Terta M, Ennaji MM. Evaluation of phytopathogenic effect of *Pectobacterium carotovorum* subs*p. carotovorum* isolated from symptomless potato tuber and soil. British Journal of Applied Science & Technology. 2014;4(1): 67-78.

28. Li X, De Boer SH. Selection of Polymerase Chain Reaction primers from an RNA intergenic spacer region for specific detection of *Clavibacter michiganensis* subsp *sepedonicus*. Phytopathol. 1995;85: 837-42.

29. Darrasse A, Priou S, Kotoujansky A, Bertheau Y. PCR and restriction fragment length polymorphism of a pel gene as a tool to identify *Erwinia carotovora* in relation to potato diseases. Appl Environ Microbiol. 1994;60:1437–43.

30. Garcia-Vallve S, Palau J, Romeu A. Horizontal gene transfer in glycosyl hydrolases inferred from codon usage in *Escherichia coli* and *Bacillus subtilis*. Mol Biol Evol. 1999;9:1125-34.

31. Eisa N, Kamaruzaman S, Zainal AMA, Ganesan V. Characterization of *Pectobacterium carotovorum* subs*p. carotovorum* as a new disease on Lettuce in Malaysia. Australasian Plant Dis Notes. 2013;8:105–7.

32. Rahman MM, Eaqub Ali M, Khan AA, Hashim U, Akanda AM, Hakim MA. Characterization and identification of soft rot bacterial pathogens in Bangladeshi potatoes. Afr J Microbiol Res. 2012;6(7): 1437-45.

33. Baghaee-Ravari S, Rahimian H, Shams-Bakhsh M, Lopez-Solanilla E, Antunez-Lamaz M, Rodriguez-Penzuela P. Characterization of *Pectobacterium* species from Iran using biochemical and molecular methods. Eur J Plant Pathol. 2011;129:413–25.

34. Nazerian E, Sijam K, Zainal Abidin MA, Vadamalai G. First report of cabbage soft rot caused by *Pectobacterium carotovorum*

subsp. *carotovorum* in Malaysia. Plant Dis. 2011;95(4):491.

35. Seo ST, Furuya N, Lim CK, Tsuchiya K, Takanami Y. Variation of *Erwinia carotovora* subsp. *carotovora* isolated from Korea. J Fac Agr Kyushu U. 2001;45:431–36.

36. Yahiaoui-Zaidi R, Jouan B, Andrivon D. Biochemical and molecular diversity among *Erwinia* isolates from potato in Algeria. Plant Pathol. 2003;52:28–40.

37. Serfontein S, Logan C, Swanepoel AE, Boelema BH, Theron DJ. A potato wilt disease in South Africa caused by *Erwinia carotovora* subspecies *carotovora* and *Erwinia chrysanthemi*. Plant Pathol. 1991; 40:382–86.

38. Costa AB, Eloy M, Cruz L, Janse JD, Oliveira H. Studies on pectolytic *Erwinia* spp. in Portugal reveal unusual strains of *E. carotovora* subsp. *atroseptica*. J Plant Pathol. 2006;88:161–6.

39. Zhu L, Xie H, Chen S, Ma R. Rapid isolation, identification and phylogenetic analysis of *Pectobacterium carotovorum* ssp. J Plant Pathol. 2010;92(2):479-83.

40. Dolzani L, Tonin E, Lagatolla C, Prandin L, Monti-Bragadin C. Identification of Acinetobacter isolates in the A. calcoaceticus-A. baumannii complex by restriction analysis of the 16S-23S rRNA intergenic-spacer sequences. J Clin Microbiol. 1995;33:1108–13.

41. Lagatolla C, Dolzani L, Tonin E, Lavenia A, Di Michele M, Monti-Bragadin C, et al. PCR ribotyping for characterizing *Salmonella* isolates of different serotypes. J Clin Microbiol. 1996;34:2440-43.

42. Navarro E, Simonet P, Normand P, Bardin R. Characterisation of natural populations of *Nitrobacter* spp. using PCR-RFLP analysis of the ribosomal intergenic spacer. Arch Microbiol. 1992;157:107–15.

43. Ma B, Hibbing ME, Kim HS, Reedy RM, Yedidia I, Breuer J, Glasner JD, Perna NT, Kelman A, Charkowski AO. Host range and molecular Phylogenies of the soft rot enterobacterial genera *Pectobacterium* and *Dickeya*. Phytopathol. 2007;97(9): 1150–63.

44. Glasner JD, Marquez-Villavicencio M, Kim HS, Jahn CE, Ma B, Biehl BS, Rissman AI, Mole B, Yi X, Yang CH. Niche-Specificity and the Variable Fraction of the

Pectobacterium Pan-Genome. Mol Plant Microbe Interact. 2008;21(12):1549–60.

45. Terta M, Azelmat S, Ait M'hand R, Achbani E, Barakate M, Ennaji MM, et al. Molecular typing of *Pectobacterium carotovorum* isolated from potato tuber soft rot in Morocco. Ann Microbiol. 2012;62(4):1411-7.

5

Effects of Exogenous Salicylic Acid on the Antioxidative System in Bean Seedling Treated with Manganese

Issam Saidi[1*], Nasreddine Yousfi[2], Wahbi Djebali[1] and Yacine Chtourou[3]

[1]Department of Biology, Laboratory of Plant Physiology and Biotechnology, Faculty of Sciences of Tunis, Campus Universitaire, 1060 Tunis, Tunisia.
[2]Department of Biology, Laboratory of Plant Extrêmophiles, Biotechnology Center of Borj Cédria, BP 901, 2050 Hammam-Lif, Tunisia.
[3]Department of Biology, Laboratory of Toxicology and Environmental Microbiology, Faculty of Sciences of Sfax, Tunisia.

Authors' contributions

This work was carried out in collaboration between all authors. Author IS designed the study, performed the statistical analysis, wrote the protocol and wrote the first draft of the manuscript. Authors NY, WD and YC managed the analyses of the study. All authors read and approved the final manuscript.

<u>Editor(s):</u>
(1) Nikolaos Labrou, Department of Agriculture Biotechnology, Agricultural University of Athens, Greece.
(2) Samuel Peña-Llopis, Oncology Division, University of Texas Southwestern Medical Center, USA.
<u>Reviewers:</u>
(1) Anonymous, Egypt.
(2) Myrene Roselyn D'souza, Department of Chemistry (P.G. Biochemistry), Mount Carmel College, India.
(3) Anonymous, Cuba.

ABSTRACT

In the present study we investigated the role of salicylic acid (SA) in regulating Mn-induced oxidative stress in bean (*Phaseolus vulgaris*) leaves. Exposure of plants to 100 µM Mn inhibited biomass production and intensively increased Mn accumulation in leaves. Concomitantly, Mn significantly enhanced protein carbonyl, H_2O_2 content and lipid peroxidation as indicated by malondialdehyde (MDA) accumulation. SA (10, 50 and 100 µM) pretreatment alleviated the negative effect of Mn on plan growth and led to decrease in oxidative stress induced by Mn stress. Furthermore, SA enhanced the activities of catalase (CAT, EC 1.11.1.6), ascorbate peroxidase (APX, EC 1.11.1.11), but lowered that of superoxide dismutase (SOD, EC 1.15.1.1) and guaiacol peroxidase (POD, EC

Corresponding author: Email: issambenahmed@yahoo.fr

1.11.1.7). The data suggest that the beneficial effect of SA could be related to avoidance of oxidative damage upon exposure to Mn thus reducing the negative consequences of oxidative stress caused by Mn toxicity.

Keywords: Manganese; salicylic acid; oxidative stress; antioxidants; Phaseolus vulgaris.

1. INTRODUCTION

Manganese (Mn) is a crucial trace metal required in numerous cellular processes, including metabolism and oxidative stress defense [1]. However, it has been reported that excess Mn disturbs the metabolism of plants and inhibit the plant growth [2]. Mn toxicity symptoms in plants appear first in shoots and these are often more sensitive toxicity parameters than vegetative growth [3]. For many species such as barley [4], bean [5], sunflower [6] and cow pea [7] first Mn toxicity symptoms are dark brown speckles, desiccation and shedding of old leaves [8]. The toxic effects of manganese can be related to the generation of reactive oxygen species (ROS), such as superoxide radicals ($O_2^{\cdot-}$), hydroxyl radical ($OH^{\cdot-}$) and hydrogen peroxide (H_2O_2) in the plants, which damage the membranes and all the macromolecules like lipids, proteins and nucleic acids in the cells [2].

Salicylic acid (SA), a naturally occurring plant hormone, has been shown to be an essential signal molecule involved in local defense reactions and induction of systemic resistance response of plants after pathogen attack [9]. The role of SA in plant tolerance to abiotic stresses such as drought [10], chilling [11], heat [12] and osmotic stress [13] has also been reported.

Salicylic acid (SA) is involved in regulating plant responses under heavy metal-induced toxicity [14-16]. Interestingly, the majority of SA-regulated heavy metal stresses in plants are involved in antioxidative responses, suggesting that SA is an internal signal molecule interacts with ROS signal pathway. In addition, SA could contribute to maintaining cellular redox homeostasis through the regulation of antioxidant enzyme activity [17] and induction of the alternative respiratory pathway [18].

Bean (*Phaseolus vulgaris*) is a widely used plant cultured throughout the word. It accumulates considerable amount of some metals, such as Cd, Cu and Zn [19]. Since Mn produces oxidative damage in higher plants, an enhancement of the antioxidant properties of bean seedling is expected as a consequence of SA supplementation under Mn stress conditions. Therefore the present study was undertaken to evaluate the toxic mechanism associated with Mn exposure and to Investigate the possible mediatory role of SA in protecting plants from Mn-induced oxidative stress.

2. MATERIALS AND METHODS

2.1 Plant Material and Growth Conditions

Bean seeds (*Phaseolus vulgaris*) were disinfected with 1% NaOCl for 5 min, then washed thoroughly with distilled water and germinated between wet paper towels at 24°C in the dark. Four days later, obtained seedlings were transferred into plastic beakers (6 L capacity, 6 plants per beaker) filled with nutrient solution containing: 1.0 mM $MgSO_4$, 2.5 mM $Ca(NO_3)_2$, 1.0 mM KH_2PO_4, 2.0 mM KNO_3, 2.0 mM NH_4Cl, 50 µM EDTA–Fe–K, 30 µM H_3BO_3, 1.0 µM $ZnSO_4$, 1.0 µM $CuSO_4$ and 30 µM $(NH4)_6Mo_7O_{24}$. After an initial growth period of 7 days in different SA concentrations (10, 50 and 100 µM), treatments were performed by adding 100 µM $MnCl_2$ to the nutrient solution. Mn dose used in this work are chosen appropriately to expose the plants from low to moderate levels of Mn. Plants were grown in a growth chamber with a 16-h-day (25°C)/8-h-night (20°C) cycle, an irradiance of 150 µmol m^{-2} s^{-1}, and 65-75% relative humidity. The nutrient solution was buffered to pH 5.5 with HCl/KOH, aerated, and changed twice per week. After 4 days of Mn-treatment, roots of the harvested plantlets were soaked in 20 mM EDTA for 15 min to remove adsorbed metals and washed carefully using distilled water to eliminate any contamination. Primary leaves were harvested and immediately stored in nitrogen liquid. Three independent repetitions of the whole experiment were performed in order to check reproducibility. For biomass production, Mn determination and biochemical analyzes five plantlets from each replication of all treatments were selected.

2.2 Determination of Ion Concentrations

Dry plant material was powdered and wet-digested in acid mixture ($HNO_3:HClO_4$, 3:1, v/v) at 100°C. Mn concentrations were estimated by atomic absorption spectrometry (Perkin-Elmer, Analyst 300) using an air–acetylene flame.

2.3 Determination of Lipid Peroxidation, Hydrogen Peroxide and Protein Carbonyl Contents

The level of lipid peroxidation in plant leaves was determined by estimation of the thiobarbituric acid (TBA) reactive substances which was expressed as the malondialdehyde (MDA) concentration based on the method of Hodges et al. [19]. Lipid peroxidation level was expressed as nmol MDA formed using an extinction coefficient of 155 mM^{-1} cm^{-1}.

Hydrogen peroxide (H_2O_2) levels were determined according to Sergiev et al. [20]. Results were expressed as nmol g^{-1} FW.

Protein carbonyls were determined using 2,4-dinitrophenylhydrazine (DNPH) and the basis of the assay involved the reaction between protein carbonyl and DNPH to form protein hydrazone [21]. Results were expressed as nmol DNPH conjugated mg^{-1} protein.

2.4 Determination of Antioxidative Enzyme Activities

Frozen leaf tissue (0.4 g) was homogenized in 4 mL ice-cold extraction buffer (50 mM potassium phosphate, pH 7.0, 0.4% PVPP) using a pre-chilled mortar and pestle. The homogenate was squeezed through a nylon mesh and centrifuged for 30 min at 14,000 g at 4°C. The supernatant was used for assays of the activities of superoxide dismutase (SOD), catalase (CAT), ascorbate peroxidase (APX) and Peroxidase (POD). All spectrophotometric analyses were conducted at 25°C.

The activity of SOD (EC1.15.1.1) was assayed by measuring its ability to inhibit the photochemical reduction of nitroblue tetrazolium (NBT) following the method of Beauchamp and Fridovich [22]. SOD activity was expressed as U mg^{-1} protein.

CAT (EC1.11.1.6) activity was assayed by the decomposition of hydrogen peroxide according to Aebi [23]. CAT activity was expressed as U mg^{-1} protein.

APX (EC1.11.1.1) activity was determined by the method of Nakano and Asada [24]. APX activity was expressed as U mg^{-1} protein.

The activity of POD (EC1.11.1.7) was determined in terms of oxidation of guaiacol by measuring increase in absorbance at 470 nm [25]. POD activity was expressed as U mg^{-1} protein.

2.5 Determination of Soluble Protein Concentration

Soluble protein concentration was measured according to Bradford [26] using the bovine serum albumin (BSA) protein assay reagent (Pierce, BSA Protein Assay Kit, USA) with BSA as the standard protein. All spectrophotometric measurements were performed by using a Perkin Elmer's LAMBDA 25/35/45 UV/Vis spectrophotometer.

2.6 Statistical Analysis

All statistical analyses were carried out with GraphPad Prism 4.02 for Windows (GraphPad Software, San Diego, CA). Significant differences between treatment effects were determined by 1-way ANOVA, followed by Tukey's post-hoc test for multiple comparisons with statistical significance of $P<0.05$. Number of replications (n) in tables/figures denotes individual plants measured for each parameter. Results were expressed as mean ± standard error of the mean (mean ± SEM).

3. RESULTS

3.1 Effects of SA Pretreatment on Seedling Growth under Mn Stress

Plant growth was negatively affected by Mn treatment (100 µM), reducing root and leaf dry matter by about 52.55% and 57.89% as compared with untreated plants (Table 1). Although SA pretreatments did not affect root dry matter of plants grown with or without Mn, they markedly alleviated Mn-induced leaf growth inhibition. Under Mn stress conditions, SA pretreatments enhanced leaf dry matter and the most significant effect was observed at 100 µM SA (increase by about 43% in leaf compared to Mn-treated plant) (Table. 1).

Table 1. Effects of SA pretreatment on root and leaf dry matter in 14-day-old bean plant submitted during 4 days to 100 µM MnCl$_2$. Data are means ± SEM (n=5). Values within rows followed by the same letter(s) are not significantly different according to Tukey's test, (P<0.05)

SA (µM)	Mn in nutrient solution	
	0 (µM)	100 (µM)
Root DM (g plant^{-1})		
0	0.137±0.009a	0.065±0.003b
10	0.147±0.011a	0.063±0.001b
50	0.140±0.017a	0.072±0.012b
100	0.142±0.012a	0.076±0.014b
Leaf DM (g plant^{-1})		
0	0.19±0.001b	0.08±0.004d
10	0.21±0.004b	0.11±0.002c
50	0.24±0.009a	0.12±0.001c
100	0.26±0.011a	0.14±0.002c

3.2 Effects of SA Pretreatment on Mn Distribution

Mn addition to the nutrient solution resulted in a high accumulation of this metal within plant organs reaching 3.14 mg g^{-1} DM in roots and a value of 4.55 mg g^{-1} DM in leaves (Table. 2). Pretreatment with SA before application of Mn markedly decreased Mn concentration in both roots and leaves. In comparison with Mn treatment, pretreatment with SA (100 µM) decreased Mn concentration by 19.11% and 29.45%, respectively in roots and leaves (Table 2).

Table 2. Effects of SA pretreatment on Mn distribution in 14-day-old bean plant submitted during 4 days to 100 µM MnCl$_2$. Data are means ± SEM (n=5). Values within rows followed by the same letter(s) are not significantly different according to Tukey's test, (P<0.05)

SA (µM)	Mn concentration (mg g^{-1} DM)	
	Roots	Leaves
0	3.14±0.11a	4.55±0.101a
10	2.95±0.14a	4.22±0.164a
50	2.82±0.16b	4.02±0.102b
100	2.54±0.08c	3.21±0.168c

3.3 Effects of SA Pretreatment on H$_2$O$_2$, MDA Concentration and Protein Carbonyl Contents in Leaves of Bean Seedlings under Mn Stress

Mn addition increased H$_2$O$_2$ concentration by more than twice as compared to the control (Table 3). SA alone did not significantly affect H$_2$O$_2$ production rate (Table 3). By contrast, its concentration in leaves of seedlings supplied with 10, 50 and 100 µM SA was decreased by

respectively 16.59%, 21.99% and 30.54% relative to Mn-stressed plants grown without SA application.

As compared to the control, Mn-treated plants exhibited a higher leaf MDA concentration (Table 3). Plant pretreatment with SA before Mn application significantly decreased MDA level in a dose-dependent manner, the effect being more pronounced with the highest applied SA (100 µM) concentrations. However, no major changes were observed in MDA level in the presence of SA alone.

Protein carbonyl (PCO) content increased upon Mn exposure by approximately 33.45% as compared to the control (Table 3). By contrast, pretreatment with SA before Mn application decreased the level of PCO by about, 12, 15.8 and 21.6% at 10, 50 and 100 µM, respectively as compared to Mn-stressed plants (Table. 3).

3.4 Effects of SA Pretreatment on Antioxidant Enzyme Activities in Leaves of Bean Seedlings under Mn Stress

Under Mn stress conditions SOD and POD activities in leaves exposed to 100 µM Mn were observed to be 41.5-38% higher than those of the control (Figs. 1A-B). By contrast, CAT and APX activities were decreased in Mn-treated leaves by about 43.5% and 40% as compared to the control (Fig. 1C-D). SA pretreatment resulted in significant decreases in SOD and POD activities upon Mn exposure and alleviated the inhibitory effect Mn on CAT and APX activities. Leaf SOD and POD activities were strongly affected by SA pretreatment, especially at 100 µM (approximately, 24% and 20% lesser than Mn alone treatment). Moreover, SA were effective in alleviating Mn-inhibited leaf APX and CAT activities, The most prominent effect was at 100 µM SA, the concentration that induced an increase of 26-30% in CAT and APX activities (Figs. 1C-D). However, no noticeable variation was observed in antioxidant enzyme activities of plants treated with SA only.

Table 3. Effects of SA pretreatment on MDA, H_2O_2 and PCO concentration in 14-day-old bean plant submitted during 4 days to 100 µM $MnCl_2$. Data are means ± SEM (n=5). Values within rows followed by the same letter(s) are not significantly different according to Tukey's test, (P<0.05)

Treatment		H_2O_2 (nmol g^{-1}FW)	MDA (nmol g^{-1}FW)	PCO (nmol DNPH mg^{-1} prot)
SA (µM)	Mn (µM)			
0	0	8.11±0.110c	5.80±0.150d	2.81±0.19c
10	0	7.48±0.019c	5.98±0.065d	2.60±0.11c
50	0	8.57±0.025c	5.02±0.144d	2.77±0.21c
100	0	8.99±0.054c	5.68±0.012d	2.86±0.14c
0	100	17.78±0.148a	11.43±0.11a	3.75±0.15a
10	100	14.83±0.108b	9.13±0.181b	3.30±0.15b
50	100	13.87±0.108b	8.24±0.101b	3.16±0.10b
100	100	12.35±0.108b	7.82±0.011c	2.94±0.21b

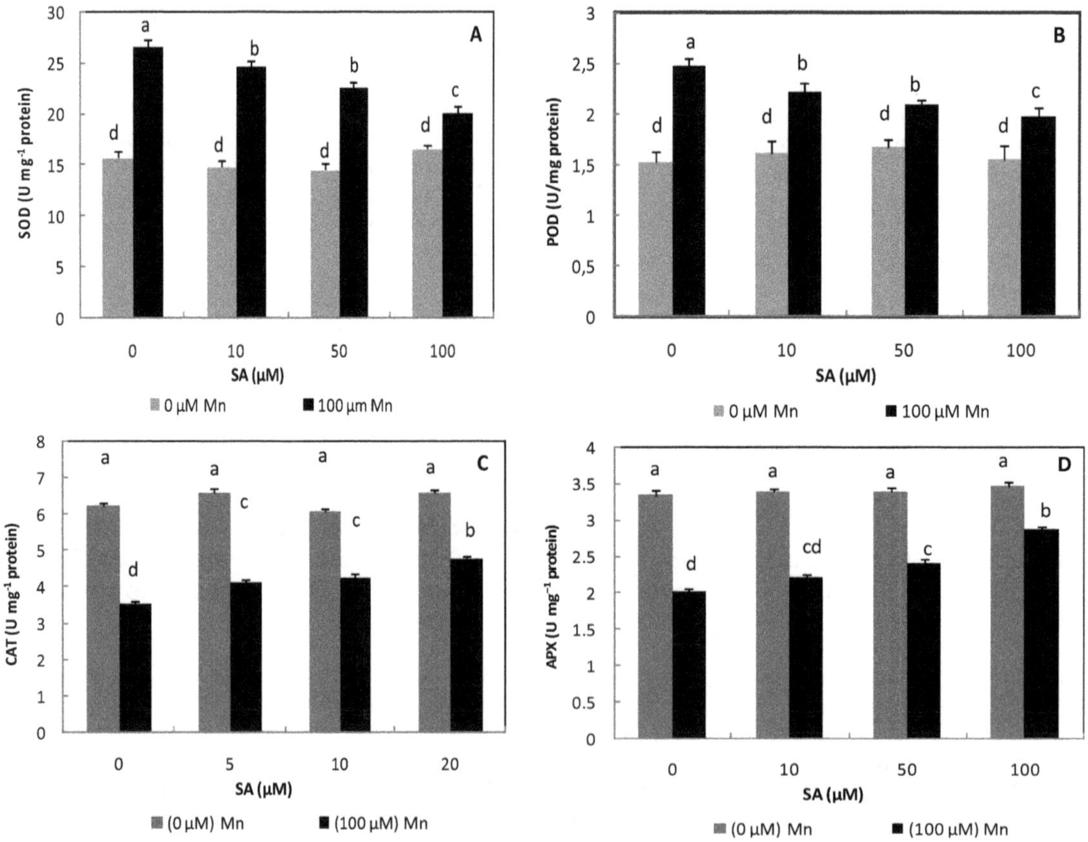

Fig. 1. Effects of SA pretreatment on SOD (A), CAT (B), APX (C) and POD (D) activities in 14-day-old bean plant submitted during 4 days to 100 µM $MnCl_2$. Data are means ± SEM (n=5). Values within rows followed by the same letter(s) are not significantly different according to Tukey's test, (P<0.05)

4. DISCUSSION

In this study, we provide evidence that salicylic acid is able to regulate Mn-induced oxidative stress in bean leaves. Pretreatment with SA before Mn application resulted in significant increase in leaf DM, when compared to Mn-stressed plants grown without SA addition. The ameliorative impact of SA on growth as observed in the present study has already been reported in different crop plants under abiotic stress conditions and this was ascribed to the role of SA in Mn distribution and nutrient uptake [27].

A variety of abiotic stresses, including heavy metals, cause molecular damage to plant cells either directly or indirectly through a burst of active oxygen species (AOS) [28]. These oxygen species (O^-, OH^-, H_2O_2) can convert fatty acids to toxic lipid peroxides, which destroy biological membranes. The present study showed that plant leaves exposed to Mn stress exhibit increased levels of H_2O_2 and MDA as a consequence of the generation of ROS. These findings are consistent with previous reports by Shi and Zhu [29] for cucumber plant, indicating that antioxidant enzymes are not a sufficient defense system.

A drastic increase in H_2O_2 may have as consequence a lower extensibility of plant cell walls, which can rapidly terminate growth [30]. This could explain the decrease of fresh mass observed in bean leaves at 100 µM of Mn. Increased of H_2O_2 content may also inactivate enzymes by oxidizing their thiol groups. However, in Mn-stressed plants pretreated with SA, H_2O_2 level and lipid peroxidation were much lower than in plants treated with Mn only. SA subdued H_2O_2 and MDA formation is supported by several recent reports [29], which indicated the protective effect of SA in lowering H_2O_2 and lipid peroxidation rate under Mn stress conditions.

SA influence on avoiding the toxic effects of Mn may be a consequence of very different primary effects connected with oxidative stress [13] and stabilization of cell membranes [31], leading to the increase of general stress tolerance. SA pretreatment led to a decrease in oxidative injuries as evidenced by decreased H_2O_2 and lipid peroxidation levels. SA may act directly as an antioxidant to scavenge the ROS and/or indirectly modulate redox balance through activation of antioxidant responses. Indeed, it is a direct scavenger of hydroxyl radical and an iron chelating compound, thereby inhibiting the direct impact of hydroxyl radicals as well as their generation *via* the Fenton reaction [32].

The data suggest that endogenous SA plays an important antioxidant role in protecting plants from oxidative stress. In SA-pretreated bean plants, the initially decreased activities of SOD and increased activities of CAT and APX cooperatively controlled the Mn-induced H_2O_2 at high homeostatic levels contrarily to the mode during plant–pathogen interactions [33]. It seems to suggest that SA-reduced H_2O_2 permit bean seedling to respond more effectively to Mn-induced oxidative damage. It is well established that CAT has a high reaction rate but a low affinity for H_2O_2, whereas APX has a high affinity for H_2O_2 and is able to detoxify low concentrations of H_2O_2 [34]. Therefore, it is possible that stimulation of CAT and APX activities by SA decreases the level of H_2O_2 in bean leaves, which may be a possible mechanism in plant defense strategy against Mn-induced oxidative stress.

Oxidative stress is caused by a serious cell imbalance between the production of ROS (H_2O_2), and antioxidative enzymes, which leads to dramatic physiological disorders. The present study has demonstrated that SA alleviates Mn-induced oxidative stress as reflected by reduced H_2O_2, and MDA content. ROS can initiate the peroxidation and destruction of lipid bilayer of cell membrane and consequently affects cell functions. Cell membranes are among the first targets of a number of plant stresses, and the maintenance of membrane integrity and stability is of major importance for stress tolerance [35,36].

The protective role of SA in Mn-Treated plants can be attributed to SA-inhibited lipid peroxidation process, contributing to membrane stability and SA-controlled Mn-induced H_2O_2 at high homeostatic levels by modulating enzymatic (SOD, APX and CAT) activity. SA might influence H_2O_2 signaling pathways in plant defense against Mn, should be further investigated to dissect the complicated network of SA and its involvement in plant defense at a molecular level.

5. CONCLUSION

Based on our results, we propose that the protective role of SA can be related to the decrease in lipid peroxidation, the improvement of scavenging capability of ROS, and the decrease in Mn uptake in plant. SA may act directly as an antioxidant to scavenge the ROS and/or indirectly modulate redox balance through activation of antioxidant responses.

COMPETING INTERESTS

Authors have declared that no competing interests exist.

REFERENCES

1. Cheton PLB, Archibald FS. Complexes de manganèse et la production et le balayage des radicaux libres hydroxyls. F R Biol Med. 1988;5:325–333

2. Lidon FC, Teixeira MG. Oxygen radical production and control in the chloroplast of Mn-treated rice. Plant Sci. 2000;152:7–15.

3. Marschner H. Mineral nutrition of higher plants. 2nd edition, Academic Press, UK. 1995;889:315-317.

4. Williams ED, Vlamis J. The effect of silicon on yield and manganese-54 uptake and distribution in the leaves of barley plants grown in culture solutions. Plant Physiol. 1957;32:404-409.

5. Horst WJ, Marschner H. Symptome von ManganÜberschuss bei Bohnen (*Phaseolus vulgaris*). Zeitscrift für. Pflanzenernaehr. Bodenkd. 1978;141:129-142.

6. Blamey FPC, Joyce DC, Edwards DG, Asher CJ. Role of trichomes in sunflower tolerance to manganese toxicity. Plant Soil. 1986;91:171-180.

7. Horst W.J. Factors responsible for genotypic manganese tolerance in cowpea (*Vigna unguiculata*). Plant Soil. 1983;72:213-218.

8. Horst WJ. The physiology of manganese toxicity. In: Graham R.D, Hannam, R.J. and Uren N.C. (Eds.), Manganese in Soils and Plants, Kluwer Academic Publishers, Dordrecht, The Netherlands. 1988;175-188.

9. Horvath E, Szalai G, Janda T. Induction of abiotic stress tolerance by salicylic acid Signaling. J Plant Growth Reg. 2007;26:290–300.

10. Munne-Bosch S, Penuelas J. Photo-and antioxidative protection, and a role for salicylic acid during drought and recovery in field-grown *Phillyrea angustifolia* plants. Planta. 2003;217:758–66.

11. Kang HM, Saltveit ME. Chilling tolerance of maize, cucumber and rice seedling leaves and roots are differentially affected by salicylic acid. Physiolgia Plantarum. 2002;115:571–76.

12. Larkindale J, Knight MR. Protection against heat stress induced oxidative damage in Arabidopsis involves calcium, abscisic acid, ethylene, and salicylic acid. Plant Physiology. 2002;128:682–95.

13. Borsani O, Valpuesta V, Botella MA. Evidence for a role of salicylic acid in the oxidative damage generated by NaCl and osmotic stress in Arabidopsis seedlings. Plant Physiology. 2001;126:1024–30.

14. Yang Y, Qi M, Mei C. Endogenous salicylic acid protects rice plants from oxidative damage caused by aging as well as biotic and abiotic stress. Plant J. 2004;40:909–19.

15. Hayat Q, Hayat S, Irfan M, Ahmad A. Effect of exogenous salicylic acid under changing environment: A review. Environmental and Experimental Botany. 2010;68:14–25.

16. Drazic G, Mihailovic N. Modification of cadmium toxicity in soybean seedlings by salicylic acid. Plant Sci. 2005;168:511–517.

17. Slaymaker DH, Navarre DA, Clark D, Del-Pozo O, Martin GB, Klessig DF. The tobacco salicylic acid-binding protein 3 (SABP3) is the chloroplast carbonic anhydrase, which exhibits antioxidant activity and plays a role in the hypersensitive defense response. Proceedings of the National Academy of Sciences of the United States of America. 2002;99:11640–45.

18. Moore AL, Albury MS, Crichton PG, Affourtit C. Function of the alternative oxidase: is it still a scavenger? Trends in Plant Science. 2002;7:478–81.

19. Chaoui A, Mazhoudi S, Ghorbal MH, El Ferjan E. Le cadmium et le zinc induction de la peroxydation des lipides et des effets sur l'activité des enzymes anti-oxydantes de haricot (*Phaseolus vulgaris* L.). Plant Science. 1997;127(2):139-147.

20. Hodges D.M, DeLong J.M, Forney C.F, Prange R.K. Improving the thiobarbituric acid-reactive-substances assay for estimating lipid peroxidation in plant tissues containing anthocyanin and other interfering compounds. Planta. 1999;207:604-11.

21. Sergiev I, Alexieva V, Karanov E. Effect of spermine, atrazine and combination between them on some endogenous

protective systems and stress markers in plants. Comptes rendus de l'Académie bulgare des Sciences. 1997;51:121–24.

22. Reznick AZ, Packer L. Oxidative damage to protein: Spectrophotometric method for carbonyl assay. Methods Enzymol. 1994;233:357-363.

23. Beauchamp C, Fridovich I. Superoxide dismutase: Improved assays and an assay applicable to acrylamide gels. Anal Biochem. 1971;44:276–87.

24. Aebi H. Catalase in vitro. Method. Enzymol. 1984;105:121-126.

25. Nakano Y, Asada K. Hydrogen peroxide is scavenged by ascorbate-specific peroxidase in spinach chloroplasts. Plant Cell Physiol. 1981;22:867-80.

26. Chance B, Maehly A.C. Assay of catalases and peroxidases. Methods in Enzymology. 1955;2:764-75.

27. Bradford MM. A rapid and sensitive method for the quantitation of microgram quantities of protein utilizing the principle of protein-dye binding. Anal Biochem. 1976;72:248–54.

28. Wang YS, Wang J, Yang ZM, Wang QY, Li B, Li SQ, Lu YP, Wang SH, Sun X. Salicylic acid modulates aluminum-induced oxidative stress in roots of Cassia tora. Acta Bot Sin. 2004;46:819–828.

29. Zhang FQ, Wang YS, Lou ZP, Dong JD. Effect of heavy metal stress on antioxidative enzymes and lipid peroxidation in leaves and roots of two mangrove plant seedlings (Kandelia candel and Bruguiera gymnorrhiza). Chemosphere. 2007;67:44–50.

30. Shi Q, Zhu Z. Effects of exogenous salicylic acid on manganese toxicity, element contents and antioxidative system in cucumber. Environ Exp Bot. 2008;63:317-326.

31. Schützendübel A, Polle A. Plant responses to abiotic stresses: Heavy metal induced oxidative stress and protection by mycorrhization. J Exp Botany. 2002;53:1351-1365.

32. Mishra A, Choudhuri M.A. Effects of salicylic acid on heavy metal induced membrane degradation mediated by lipoxigenase in rice. Biologia Plantarum. 1999;43:409–15.

33. Dinis TC, Maderia VM, Almeida LM. Action of phenolic derivates (acetaminophen, salicylate, and 5-aminosalicylate) as inhibitors of membrane lipid peroxidation and as peroxyl radical scavengers. Archives of Biochem and Biophysics. 1994;315:161–69.

34. Chen Z, Silva H, Klessig DF. Active oxygen species in the induction of plant systematic acquired resistance by salicylic acid. Science. 1993;262:1883–86.

35. Mittler R. Oxidative stress, antioxidants and stress tolerance. Trends in Plant Science. 2002;7:405–41.

36. Djanaguiraman M, Devi D.D, Shanker A.K, Sheeba J.A, Bangarusamy U. Selenium an antioxidative protectant in soybean during senescence. Plant Soil. 2005;272:77-86.

6

Genotyping and Nucleotide Sequences of Growth Hormone Releasing Hormone and Its Receptor Genes in Egyptian Buffalo

Othman E. Othman[1*], Mohamed F. Abdel-Samad[1], Nadia A. Abo El-Maaty[1] and Karima M. Sewify[2]

[1]Department of Cell Biology, National Research Center, Dokki, Egypt.
[2]Department of Zoology, Girl Faculty, Ain Shams University, Egypt.

Authors' contributions

This work was carried out in collaboration between all authors. Author OEO designed the study, followed up the practical work and wrote the final version of the manuscript. Author MFAS managed the analyses of the study, managed the literature searches and wrote the first draft of the manuscript. Author NAAEM performed the practical work. Author KMS followed up the steps of the search. All authors read and approved the final manuscript

Editor(s):
(1) Giuliana Napolitano, Department of Biology (DB), University of Naples Federico II, Naples, Italy.
Reviewers:
(1) Anonymous, Syiah Kuala University, Indonesia.
(2) Anonymous, University of Studies of Bari "Aldo Moro", Italy.
(3) Anonymous, University of Hong Kong, China.
(4) Anonymous, Islamic State Riau University, Indonesia.

ABSTRACT

Aim: The hypothalamic hormone, growth hormone-releasing hormone, is the principal stimulator of pituitary growth hormone (*GH*) synthesis and secretion. *GHRH* and its receptor (*GHRHR*) provide important functions in the regulation of the *GH* axis and in the development and proliferation of pituitary somatotropic axis. This study aimed to identify the genotypes and nucleotide sequences of two multifunctional genes; growth hormone-releasing hormone (*GHRH*) and its receptor (*GHRHR*) in Egyptian buffalo.

Methodology: Genomic DNA was extracted from blood samples of 100 healthy buffaloes maintained at the Mahlet Mussa and El-Gmeasa herds from 2010 to 2012. PCR was performed using primers flanking a 296-bp fragment from *GHRH* gene and a 425-bp fragment from *GHRHR* gene of Egyptian buffalo. The PCR-amplified fragments were digested with *HaeIII* (*GHRH*) and

Corresponding author: Email: othmanmah@yahoo.com

*Eco*57I (*GHRHR*), electrophoresed and analyzed on agarose gels stained with ethidium bromide. The two amplified fragments were also sequenced and aligned with published sequences.

Results: Depending on the presence of the restriction site at 241^242 position (GG^CC) in 296-bp amplified fragments of *GHRH,* we genotyped all tested buffalo animals as AA. Due to the absence of the restriction site at position 300^301 ([CTGAAG(N)$_{16}$^]) in the amplified fragment of *GHRHR* (425-bp), we genotyped the tested animals as AA. The Egyptian buffalo *GHRH* and *GHRHR* nucleotide sequences were submitted to NCBI/Bankit/GenBank and have the accession numbers JN967799 and KC295414, respectively.

Conclusion: The Egyptian buffaloes are characterized by best production traits like high milk fat content as well as higher average daily gain and body weight where they are possess with fixed *GHRH*^{AA} and *GHRHR*^{AA} genotypes whlch were deslred genotypes for mllk and growth production traits in different cattle breeds and the cattle are genetically homologous with buffaloes. To the best of our knowledge, these polymorphic sites are not identified in other buffalo populations. The identification of genotypes and nucleotide sequences of these two multifunctional genes may be useful in future marker-assisted selection (MAS) for more efficient breeding and genetic conservation programs of Egyptian buffalo.

Keywords: Buffalo; GHRH; GHRHR; PCR; RFLP.

1. INTRODUCTION

The great adaptive capacity of Egyptian buffaloes (*Bubalus bubalis*) to tropical climates and excellent nutritional efficiency, resistance to diseases, together with the good productive and reproductive potential make these animals one of the main sources for milk and meat in Egypt. The improvement of livestock productivity has been dependent on genetic markers that are associated with economically important productivity traits to promote more efficient selection through marker-assisted selection. Among the putative candidate markers, the genes which are related to the somatotropic axis [1,2,3,4].

The hypothalamic hormone, growth hormone-releasing hormone (*GHRH*), is the principal stimulator of pituitary growth hormone (*GH*) synthesis and secretion. Its pituitary receptor is well characterized as a member of the superfamily of G protein-coupled receptors [5]. *GHRH* and its receptor provide important functions in the regulation of the *GH* axis and in the development and proliferation of pituitary somatotropes [6].

The association between *GHRH* and an increased milk yield was confirmed by Hashizume et al. [7]. Baile and Buonomo [8] found that administering this hormone increased the metabolic activity of mammary gland cells. Furthermore, Zhao et al. [9] reported that administering *GHRH* had a significant effect on glucose transporter gene expression in the mammary gland, resulting in an increased milk

yield. Studies of Lappiera et al. [10] proved that administering recombined human *GHRH* to cow resulted in an increased milk yield as well as protein and fat content in milk.

Other studies showed that somatotropes with their synthetic equivalents increased milk production in dairy cows [11] and meat cows [12] as well as improved cattle growth rate. Moreover, Ciampani et al. [13] confirmed the role of *GHRH* in the *FSH* secretion process and thus indirectly stimulates steroidogenesis in Leydig cells and the activity of *FSH* in Sertoli cells in males.

By now, genotypes of Egyptian buffalo *GHRH* and *GHRHR* were not reported, so this study aimed to identify the genotypes and nucleotide sequences of these two multifunctional genes in Egyptian buffalo.

2. MATERIALS AND METHODS

2.1 Genomic DNA Extraction

Genomic DNA was extracted from the whole blood of 100 unrelated Egyptian buffaloes - maintained at the Mahlet Mussa and El-Gmeasa herds from 2010 to 2012- according to established protocol [14] with minor modifications. Briefly, 10 ml of blood taken on EDTA were mixed with 25 ml of cold 2X Sucrose-Triton and 15 ml double distilled water. The tubes were placed on ice for 10 min and mixed by inversion several times. After centrifugation, at 5000 rpm for 15 min at 4ºC, the pellet was re-suspended by 3 ml of nucleic lysis buffer. The content was mixed with 108 µl of 20% SDS and

150 µl of Proteinase K. The tubes were placed in a water bath at 37ºC overnight.

After the incubation, the tube contents were transferred to a 15-ml polypropylene tube and 2 ml of saturated NaCl was added and shaken vigorously for 15 sec. After centrifuging at 3500 rpm for 15 min at 4ºC, the supernatant was transferred to a clean 15-ml polypropylene tube and mixed with absolute ethanol. The tubes were agitated gently to mix the liquids and a fluffy white ball of DNA was formed. The precipitated DNA was picked up using the heat sealed pasture pipette, then washed twice in 70% ethanol and exposed to air to dry completely.

The DNA was dissolved in 200 µl TE buffer in 1.5-ml Microfuge tube and kept overnight in an incubator at 37ºC. DNA concentration was determined and diluted to the working concentration of 50 ng/µl, which is suitable for polymerase chain reaction using NanoDrop1000 Thermo Scientific spectrophotometer.

2.2 Polymerase Chain Reaction (PCR)

A PCR cocktail consisted of 1.0 µM upper and lower primers (specific for each tested gene (Table 1), 0.2 mM dNTPs and 1.25 units of Taq polymerase. The cocktail was aliquot into PCR tubes with 100 ng of buffalo DNA. The reaction was cycled for 1 min. at 94°C, 1 min at an optimized annealing temperature that was determined for each primer (Table 1) and 1 min. at 72°C for 30 cycles. The PCR products were electrophoresed on 2% agarose gel stained with ethidium bromide to test the amplification success.

2.3 Restriction Fragment Length Polymorphism (RFLP)

The PCR products for the two tested gene were digested with specific restriction enzyme for each gene (Table 1). The restriction mixture for each sample was prepared by adding 2.5 µl of 10×restriction buffer to 10 units of the appropriate restriction enzyme and the volume was completed to 5 µl by sterile water. This restriction mixture was mixed with PCR product (~25 µl) and incubated overnight at the optimum temperature of the maximum activity for each restriction enzyme. The digested PCR products were electrophoresed on a 3% agarose gel staining with ethidium bromide to detect the different genotypes of the two tested genes.

2.4 Sequence Analysis

The PCR products of each tested gene were purified and sequenced by Macrogen Incorporation (Seoul, Korea). Sequence analysis and alignment were carried out using NCBI/BLAST/blastn suite. Results of endouclease restriction were carried out using FastPCR [15]. The nucleotide sequences of the two tested genes in Egyptian buffalo were submitted to GenBank (NCBI, BankIt).

3. RESULTS AND DISCUSSSION

3.1 *GHRH* Gene

Growth hormone releasing hormone (*GHRH*) is a hypothalamic hormone which stimulates both synthesis and secretion of pituitary growth hormone (*GH*) binds to specific receptors on somatotrophs [17]. Bovine *GHRH* increased the serum concentration of endogenous *GH* [18] and increased milk production [10].

Table 1. The sequences and information of primers used in this study

Gene	Primer sequence 5' -------- 3'	PCR conditions (30 cycles)	PCR product size	Restriction enzyme used	References
GHRH	TTC CCA AGC CTC TCA GGT AA GCG TAC CGT GGA ATC CTA GT	94°C 1 min 60°C 1 min 72°C 1 min	296 bp	*Hae*III	[3]
GHRHR	ACG CCA CCC TCT TTC ACC AG CAT CCT GGG TGC TTC TTG AAG	94°C 1 min 55°C 1 min 72°C 1 min	425 bp	*Eco*57I	[16]

GHRH gene was linked to CSSM30 on bovine chromosome 13 [19] and consists of five exons separated by four introns [20].

The primers used in this study flanked a 296-bp fragment consisting of 14 base pairs from exon 2, 265 base pairs from intron 2 and 17 base pairs from exon 3 of Egyptian buffalo GHRH gene. The amplified fragments obtained from all tested buffalo DNA (100 animals) gave the expected fragment at 296-bp (Fig. 1)

Two-way sequence analysis of the GHRH amplified PCR product of buffalo DNA was conducted, using both the forward and reverse primers. The buffalo amplicon obtained was found to be 296-bp (Fig. 2). The Egyptian buffalo GHRH nucleotide sequence was submitted to nucleotide sequences database NCBI/Bankit/GenBank and has the accession number JN967799.

The sequence alignment of Egyptian buffalo GHRH with published sequence (accession number: DQ064594; Bubalus bubalis) was carried out using BLAST and showed that the Egyptian buffalo possess identities at 99% with only one gap between positions 259 and 260 and one SNP (T/C) at position 285 of our sequence (Fig. 3).

These PCR amplified fragments (296-bp) were digested with HaeIII endonuclease. Depending on the presence or absence of the restriction site at 241^242 position (GG^CC) in these amplified fragments, we can easily differentiate between 3 different genotypes: AA with two digested fragments at 241-and 55-bp, BB with three digested fragments at 193-, 55-and 48-bp and AB with four digested fragments at 241-, 193-, 55- and 48-bp.

All buffalo animals investigated in this study are genotyped as **AA** where all tested buffalo DNA amplified fragments were digested with HaeIII endonuclease and gave two digested fragments at 241- and 55-bp (Fig. 4) due to the presence of the restriction site at position 241^242 (GG^CC) (Fig. 5).

Dybus and Grzesiak [21] evaluated the relationship between the polymorphism of the GHRH and milk production traits of Polish Black-and-White. A PCR-RFLP method was used for its genotyping. The frequencies of the genotypes and alleles were as follows: 0.0545 for AA, 0.3133 for AB and 0.6322 for BB, and 0.2111 for $GHRH^A$ and 0.7889 for $GHRH^B$. There were no significant associations between GHRH/HaeIII polymorphism and milk production traits of the analyzed cows.

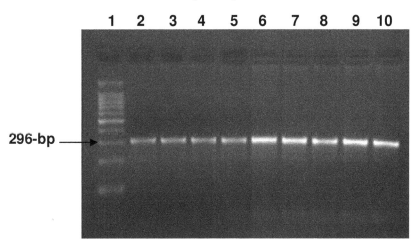

Fig. 1. Ethidium bromide-stained gel of PCR products representing amplification of GHRH gene in Egyptian buffalo
Lane 1: 100-bp ladder marker
Lanes 2-10: 296-bp PCR products amplified from Egyptian buffalo DNA

TCCCAGCCTCTCAGGTAAGCAGTTCTGACAAGAGAAGCAAGCGAGGCACTTTGAGGATG
CAGACTCGAGCTGGTCCCCAGCTGGCTCCTCAGGCAGCCTCCCTTGCTCATCTCTGGGA
GGGAGGCAGACTAAGCCCCAGAGAGGTCACCACCCAGCCCTGGTTCCAGCCCTCTCTGG
GGACGAGCAGGGCAAGAGGCGACAGAAAGACCTCACAGAGACCAAGTGAGCACAGTCCC
CTGGGCCTCCCACCCCACCCTTTGACCTCTGACTCCTTCT**ACTAGGATTCCACGGTACGC**

Fig. 2. The nucleotide sequence of Egyptian buffalo *GHRH* amplified fragment.
Forward and reverse primers with bold

```
Query 1    TCCCAGCCTCTCAGGTAAGCAGTTCTGACAAGAGAAGCAAGCGAGGCACTTTGAGGATGC   60
           ||||||||||||||||||||||||||||||||||||||||||||||||||||||||||||
Sbjct 73   TCCCAGCCTCTCAGGTAAGCAGTTCTGACAAGAGAAGCAAGCGAGGCACTTTGAGGATGC   132

Query 61   AGACTCGAGCTGGTCCCCAGCTGGCTCCTCAGGCAGCCTCCCTTGCTCATCTCTGGGAGG   120
           ||||||||||||||||||||||||||||||||||||||||||||||||||||||||||||
Sbjct 133  AGACTCGAGCTGGTCCCCAGCTGGCTCCTCAGGCAGCCTCCCTTGCTCATCTCTGGGAGG   192

Query 121  GAGGCAGACTAAGCCCCAGAGAGGTCACCACCCAGCCCTGGTTCCAGCCCTCTCTGGGGA   180
           ||||||||||||||||||||||||||||||||||||||||||||||||||||||||||||
Sbjct 193  GAGGCAGACTAAGCCCCAGAGAGGTCACCACCCAGCCCTGGTTCCAGCCCTCTCTGGGGA   252

Query 181  CGAGCAGGGCAAGAGGCGACAGAAAGACCTCACAGAGACCAAGTGAGCACAGTCCCCTGG   240
           ||||||||||||||||||||||||||||||||||||||||||||||||||||||||||||
Sbjct 253  CGAGCAGGGCAAGAGGCGACAGAAAGACCTCACAGAGACCAAGTGAGCACAGTCCCCTGG   312

Query 241  GCCTCCCACCCCACCCTTT-GACCTCTGACTCCTTCTACTAGGATTCCACGGTACGC     296
           |||||||||||||||||||| ||||||||||||||||||||||||| |||||||||||
Sbjct 313  GCCTCCCACCCCACCCTTTTGACCTCTGACTCCTTCTACTAGGATCCCACGGTACGC     369
```

Fig. 3. Sequence alignment of Egyptian buffalo *GHRH* with published sequence.
(-/T) gap and (T/C) single nucleotide polymorphism with bold

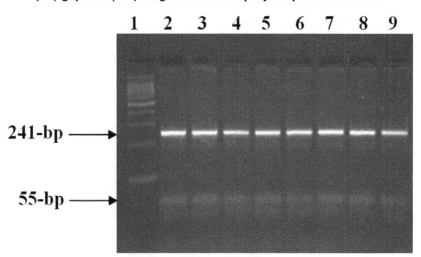

Fig. 4. The electrophoretic pattern obtained after digestion of PCR
amplified buffalo *GHRH* with *Hae*III restriction enzyme
Lane 1: 100-bp ladder marker
Lanes 2-9: Homozygous AA genotypes showed two restricted fragments at 241- and 55-bp

TCCCAGCCTCTCAGGTAAGCAGTTCTGACAAGAGAAGCAAGCGAGGCACTTTGAGGATGC
AGACTCGAGCTGGTCCCCAGCTGGCTCCTCAGGCAGCCTCCCTTGCTCATCTCTGGGAG
GGAGGCAGACTAAGCCCCAGAGAGGTCACCACCCAGCCCTGGTTCCAGCCCTCTCTGGG
GACGAGCAGGGCAAGAGGCGACAGAAAGACCTCACAGAGACCAAGTGAGCACAGTCCCC
TG**GG^CC**TCCCACCCCACCCTTTGACCTCTGACTCCTTCTACTAGGATTCCACGGTACGC

Fig. 5. Endonuclease restriction of Egyptian buffalo *GHRH* using FastPCR
GG^CC restriction site with bold

By direct DNA sequencing in 24 unrelated Korean cattle, Cheong et al. [22] identified 12 single nucleotide polymorphisms within *GHRH* gene. Among them, six polymorphic sites were selected for genotyping in beef cattle and five marker haplotypes were identified. Statistical analysis revealed that -4241A>T showed significant associations with cold carcass weight (CW) and longissimus muscle area (EMA).

Also the polymorphism of cattle *GHRH* gene using PCR-RFLP technique with *Hae*III restriction enzyme was studied by Kmiec et al. [23]. They detected two alleles *GHRH*^{A} with frequency of 28.1% and *GHRH*^{B} with frequency of 71.9%. This study proved the existence of *GHRH*/*Hae*III polymorphism in the selected gene sequence and revealed statistically higher values for the analyzed milk production traits in cows with *GHRH*^{A}/*GHRH*^{A} genotype.

Szatkowska et al. [24] analyzed the association between the *GHRH*/*Hae*III gene polymorphism with milk production traits of Polish Holstein and Jersey cows. The frequencies of genotypes and alleles for the Polish Holstein cows were 0.078 for AA, 0.339 for AB and 0.583 for BB. In all lactations, the Jersey cows with AA genotype exhibited the highest milk fat content. In the 2nd and 3rd lactations the AA Jersey cows had lower milk yields compared with the AB or BB cows.

The association of the *GHRH* gene with growth traits in Chinese native cattle was investigated by Zhang et al. [25]. PCR-SSCP and sequencing were used to detect mutations of the *GHRH* gene. One novel mutation 4251nt (C > T) was found and the frequencies of C allele were 0.8778 and 0.8476 for Qinchuan and Nanyang cattle, respectively. Body weight with the CT genotype was significantly higher than those with CC genotype in Nanyang cattle.

3.2 GHRHR Gene

The hypothalamic hormone, growth hormone-releasing hormone (*GHRH*), is the principal stimulator of pituitary growth hormone (*GH*) synthesis and secretion. Its pituitary receptor is well characterized as a member of the superfamily of G protein-coupled receptors [5]. *GHRH* and its receptor provide important functions in the regulation of the *GH* axis and in the development and proliferation of pituitary somatotropes [6].

The primers used in this study flanked a 425-bp fragment consisting of 50-bp from exon 6 and 375-bp from intron 6 of Egyptian buffalo *GHRHR* gene. The amplified fragments obtained from all tested buffalo DNA (100 animals) at 425-bp (Fig. 6).

Two-way sequence analysis of the *GHRHR* amplified PCR product of buffalo DNA was conducted, using both the forward and reverse primers. The buffalo amplicon obtained was found to be 425-bp (Fig. 7). The Egyptian buffalo *GHRHR* nucleotide sequence was submitted to nucleotide sequences database NCBI/ Bankit/GenBank and has the accession number KC295414.

The sequence alignment of Egyptian buffalo *GHRHR* with published sequence (accession number: EF600712.1; *Bubalus bubalis*) was carried out using BLAST and showed that the Egyptian buffalo possess identities at 100% with published sequence without any SNP in this shared fragment (Fig. 8).

These PCR amplified fragments (425-bp) were digested with *Eco*57I endonuclease. Depending on the presence or absence of the restriction site at position 300^301 ([CTGAAG(N)$_{16}$^], we can easily differentiate between 3 different genotypes: AA with undigested one fragment at 425-bp, BB with two digested fragments at 300- and 125-bp and AB with three digested fragments at 425-, 300- and 125-bp.

All buffalo animals investigated in this study are genotyped as **AA** where all tested buffalo DNA amplified fragments were treated with *Eco*57I endonuclease and gave one undigested fragment at 425-bp (Fig. 9) due to the absence of the restriction site at position 300^301 ([CTGAAG(N)$_{16}$^].

A RFLP was identified within a PCR amplification product of the bovine growth hormone releasing hormone receptor (*GHRHR*) gene using the restriction endonuclease *Eco*57I [16]. Digestion of the 425-bp product with *Eco*57I revealed a polymorphism with two alleles characterized by an uncut band of 425 bp (Allele A) and two cut bands of 125 and 300 bp (Allele B). Frequency of the A allele was 0.15 in the MARC (Meat Animal Research Center) reference families.

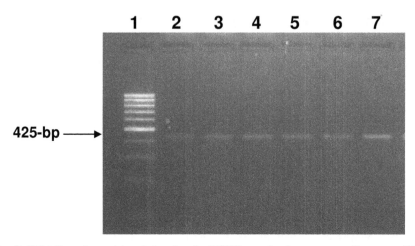

**Fig. 6. Ethidium bromide-stained gel of PCR products representing amplification of *GHRHR*
gene in Egyptian buffalo**
Lane 1: 100-bp ladder marker
Lanes 2-7: 425-bp PCR products amplified from Egyptian buffalo DNA

ACGCCACCCTCTTTCACCGGGAGAACACGGACCACTGCAGCTTCTCCACTGTAACAGTCA
TGGGTGGGGGTGCTGGTGCGGGCGAGGAGGTTGGATTAGAGATGTCAGCCTGTCCAGTCC
AGTGGGCTGACCCCGGGGCTCTGGCTTTGCCAAGGACAGAGCTGGAAAGCCCCCCTCCCC
CTTCCCGCCCCTCCTTGGGGTCAAGTCCTAAATCCTCCTGTGCCCAGCCCCGTCATTCCCT
GACTCCACTCTCTGCTCCATGTTCTGTATTCTGGTTTCATTCCCAGCCTGTAGCCCAGCCCA
GAGCACACTTCACTCCACTCTTGCTTCCATCTCAAACTTCCTCTGGGCTCTGTCTCTGCTGG
GTGTGGGTGTACCAGGCACTGGACAAAGCCAGGTCT**CTTCTTCAAGAAGCACCCAGGATG**

**Fig. 7. The nucleotide sequence of Egyptian buffalo *GHRHR* amplified fragment.
Forward and reverse primers with bold**

```
Query    1      ACGCCACCCTCTTTCACCGGGAGAACACGGACCACTGCAGCTTCTCCACTGTAACAGTCA    60
                ||||||||||||||||||||||||||||||||||||||||||||||||||||||||||||
Sbjct    7647   ACGCCACCCTCTTTCACCGGGAGAACACGGACCACTGCAGCTTCTCCACTGTAACAGTCA    7706

Query    61     TGGGTGGGGGTGCTGGTGCGGGCGAGGAGGTTGGATTAGAGATGTCAGCCTGTCCAGTCC    120
                ||||||||||||||||||||||||||||||||||||||||||||||||||||||||||||
Sbjct    7707   TGGGTGGGGGTGCTGGTGCGGGCGAGGAGGTTGGATTAGAGATGTCAGCCTGTCCAGTCC    7766

Query    121    AGTGGGCTGACCCCGGGGCTCTGGCTTTGCCAAGGACAGAGCTGGAAAGCCCCCCTCCCC    180
                ||||||||||||||||||||||||||||||||||||||||||||||||||||||||||||
Sbjct    7767   AGTGGGCTGACCCCGGGGCTCTGGCTTTGCCAAGGACAGAGCTGGAAAGCCCCCCTCCCC    7826

Query    181    CTTCCCGCCCCTCCTTGGGGTCAAGTCCTAAATCCTCCTGTGCCCAGCCCCGTCATTCCC    240
                ||||||||||||||||||||||||||||||||||||||||||||||||||||||||||||
Sbjct    7827   CTTCCCGCCCCTCCTTGGGGTCAAGTCCTAAATCCTCCTGTGCCCAGCCCCGTCATTCCC    7886

Query    241    TGACTCCACTCTCTGCTCCATGTTCTGTATTCTGGTTTCATTCCCAGCCTGTAGCCCAGC    300
                ||||||||||||||||||||||||||||||||||||||||||||||||||||||||||||
Sbjct    7887   TGACTCCACTCTCTGCTCCATGTTCTGTATTCTGGTTTCATTCCCAGCCTGTAGCCCAGC    7946

Query    301    CCAGAGCACACTTCACTCCACTCTTGCTTCCATCTCAAACTTCCTCTGGGCTCTGTCTCT    360
                ||||||||||||||||||||||||||||||||||||||||||||||||||||||||||||
Sbjct    7947   CCAGAGCACACTTCACTCCACTCTTGCTTCCATCTCAAACTTCCTCTGGGCTCTGTCTCT    8006

Query    361    GCTGGGTGTGGGTGTACCAGGCACTGGACAAAGCCAGGTCTCTTCTTCAAGAAGCACCCA    420
                ||||||||||||||||||||||||||||||||||||||||||||||||||||||||||||
Sbjct    8007   GCTGGGTGTGGGTGTACCAGGCACTGGACAAAGCCAGGTCTCTTCTTCAAGAAGCACCCA    8066

Query    421    GGATG    425
                |||||
Sbjct    8067   GGATG    8071
```

Fig. 8. Sequence alignment of Egyptian buffalo *GHRHR* with published sequence

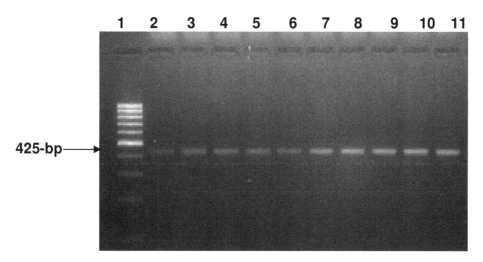

Fig. 9. The electrophoretic pattern obtained after digestion of PCR amplified buffalo *GHRHR* with *Eco*57I restriction enzyme
Lane 1: 100-bp ladder marker
Lanes 2-11: Homozygous AA genotypes showed one undigested fragment at 425-bp

Zhang et al. [26] screened the 5' flanking region, the coding region and partially introns of *GHRHR* to detect the SNPs in the predominant cattle breeds of China. The genotypes were named AA, AB and BB. Fixed effects of genotype and age were included as independent variables in the linear model. The result indicated that three linked mutations in *GHRHR* gene were significantly associated with body weight of 12 months and average daily gain of 12 months (P<0.05). The individuals with genotype AA had higher average daily gain and body weight than individuals with genotype AB. While the differences between the individuals with genotype BB and the individuals with genotype AA and AB were not significant. So, three linked mutations in *GHRHR* gene have effect on growth traits in bovine. This result proved the *GHRHR* gene as an important candidate gene controlling growth performance and carcass traits in farm animals.

4. CONCLUSION

It is concluded that the Egyptian buffaloes are characterized by best production traits like high milk fat content as well as higher average daily gain and body weight where they are possess with fixed *GHRH*^{AA} and *GHRHR*^{AA} genotypes which were reported as desired genotypes for milk and growth production traits in different cattle breeds and the cattle are genetically homologous with buffaloes. To the best of our knowledge, these polymorphic sites are not identified in other buffalo populations. The

identification of genotypes and nucleotide sequences of these two multifunctional genes may be useful in future marker-assisted selection (MAS) for more efficient breeding and genetic conservation programs of Egyptian buffalo.

COMPETING INTERESTS

Authors have declared that no competing interests exist.

REFERENCES

1. Renaville R, Gengler N, Vrech E, Prandi A, Massart S, Corradini C, Bertozzi C, Mortiaux F, Burny A, Portetellem D. Pit-1 gene polymorphism, milk yield, and conformation traits for Italian Holstein-Friesian bulls. J. Dairy Sci. 1997;80:3431-3438.

2. Sorensen P, Grochowska R, Holm L, Henryon M, Lovendahl P. Polymorphism in the bovine growth hormone gene affects endocrine release in dairy calves. J. Dairy Sci. 2002;85:1887-1893.

3. Dybus A, Kmiec M, Sobek Z, Pietrzyk W, Wisniewki B. Associations between polymorphisms of growth hormone releasing hormone (GHRH) and pituitary transcription factor 1 (PIT1) genes and production traits of limousine cattle. Arch. Tierz. Dummerstorf. 2003;46:527-534.

4. Dybus A, Grzesiak W, Kamieniecki H, Szatkowska I, Sobek Z, Blaszczyk P, Czerniawska-Piatkowska E, Zych S, Muszynska M. Association of genetic variants of bovine prolactin with milk production traits of black-and-white and Jersey cattle. Arch. Tierz. Dummerstorf. 2005;48:149-156.

5. Mayo KE. Molecular cloning and expression of a pituitary-specific receptor for growth hormone-releasing hormone. Mol. Endoc. 1992;6:1734-1744.

6. Lin-Su K, Wajnrajch MP. Growth Hormone Releasing Hormone (GHRH) and the GHRH Receptor. Rev. End. Met. Dis. 2002;3:313-323.

7. Hashizume T, Yanagimoto M, Kainuma S, Nagano R, Moriwaki K, Ohtsuki K, Sasaki K, Masuda H, Hirata T. Effects of new growth hormone-releasing peptide (KP102) on the release of growth hormone in vitro and in vivo in cattle. Anim. Sci. Technol. (Jpn). 1997;68:450-458.

8. Baile CA, Buonomo FC. Growth hormone-releasing factor effects on pituitary function, growth and lactation. J. Dairy Sci. 1987;70:467-473.

9. Zhao FQ, Moseley WM, Tucker HA, Kennelly JJ. Regulation of glucose transporter gene expression in mammary gland, muscle, and fat of lactating cows by administration of bovine growth hormone-releasing factor. J. Anim. Sci. 1996;74:183-189.

10. Lapierre H, Pelletier G, Petitclerc D, Dubreuil P, Morisset J, Gaudreau P, Couture Y, Brazeau P. Effect of human growth hormone-releasing factor (1-29)NH$_2$ on growth hormone release and milk production in dairy cows. J. Dairy Sci. 1988;71:92-98.

11. Bonneau M, Laarveld B. Biotechnology in animal nutrition, physiology and health. Livestock Prod, Sci. 1999;59:223-241.

12. Achtung TT, Buchanan DS, Lents CA, Barao SM, Dahl GE. Growth hormone response to growth hormone-releasing hormone in beef cows divergently selected for milk production. J. Anim. Sci. 2001;79:1295-1300.

13. Ciampani T, Fabbri A, Isidori A, Dufau ML. Growth hormone-releasing hormone is produced by rat Leydig cell in culture and acts as a positive regulator of Leydig cell function. Endocrinology. 1992;131:2785-2792.

14. Miller SA, Dykes DD, Polesky HF. A simple salting out procedure for extracting DNA from human nucleated cells, Nucleic Acids Res. 1988;16:1215.

15. Kalendar R, Lee D, Schulman AH. Java web tools for PCR, in silico PCR, and oligonucleotide assembly and analysis. Genomics. 2011;98(2):137-144.

16. Connor EE, Ashwell MS, Kappes SM, Dahl GE. Mapping of the bovine growth hormone-releasing hormone receptor (GHRH-R) geneto chromosome 4 by linkage analysis using a novel PCR-RFLP. J. Anim. Sci. 1999;77:793-794.

17. Frohman LA, Bowns TR, Chomczynski P. Regulation of growth hormone secretion. Frntiers in Neuroendocrinol. 1992;13:344-405.

18. Lovendahl P, Woolliams JA, Sinnett-Smith PA. Response of growth hormone to various doses of growth hormone releasing factor and thyrotropin releasing hormone administered separately and in combination to dairy calves. Can. J. Anim. Sci. 1991;71:1045-1052.

19. Barendse W, Armitage SM, Kossarek LM. A genetic linkage map of the bovine genome. Nature Genet. 1994;6:227-235.

20. Zhou P, Kazmer GW, Yang X. Bos taurus growth hormone releasing hormone gene, complete cds. GenBank, AF 242855; 2000.

21. Dybus A, Grzesiak W. GHRH/HaeIII gene polymorphism and its associations with milk production traits in Polish Black-and-White cattle. Arch. Tierz. Dummerstorf. 2006;49(5):434-438.

22. Cheong HS, Yoon DH, Kim LH, Park BL, Choi YH, Chung ER, Cho YM, Park EW, Cheong C, Oh SJ, Yi SG, Park T, Shin HD. Growth Hormone-Releasing Hormone (GHRH) polymorphisms associated with carcass traits of meat in Korean cattle. BMC Genetics. 2006;7:35-40.

23. Kmiec M, Kowalewska-Luczak I, Kulig H, Terman A, Wierzbicki H, Lepczynski A. Associations between GHRH/HaeIII restriction polymorphism and milk production traits in a herd of dairy cattle. J. Anim. Vet. Adv. 2007;6(11):1298-1303.

24. Szatkowska I, Dybus A, Grzesiak W, Jedrzejczak M, Muszyńska M. Association between the Growth Hormone Releasing Hormone (GHRH) gene polymorphism and

milk production traits of dairy cattle. J. Appl. Anim. Res. 2009;36:119–123.

25. Zhang B, Zhao G, Lan X, Lei C, Zhang C, Chen H. Polymorphism in GHRH gene and its association with growth traits in Chinese native cattle. Res. Vet. Sci. 2012;92(2): 243–246.

26. Zhang C, Chen H, Zhang L, Zhao M, Guo Y. Association of polymorphisms of the GHRHR gene with growth traits in cattle. Arch. Tierz. Dummerstorf. 2008;51(3):300-301.

7

Characterization of New Quality Protein Maize (QPM) Varieties from a Breeding Program: Analysis of Amino Acid Profiles and Development of a Variety-Diagnostic Molecular Marker

K. K. Nkongolo[1,2*], G. Daniel[1,3], K. Mbuya[4], P. Michael[2] and G. Theriault[2]

[1]Department of Biology, Laurentian University, Sudbury, Ontario, P3E 2C6, Canada,
[2]Biomolecular Sciences Program, Laurentian University, Sudbury, Ontario, P3E 2C6, Canada.
[3]Department of Chemistry and Biochemistry, Laurentian University, Sudbury, Ontario, P3E 2C6, Canada.
[4]National Maize Program, National Institute for Agronomic Study and Research, (INERA) B.P. 2037, Kinshasa 1, DR-Congo.

Authors' contributions

This work was carried out in collaboration between all authors. Author KKN designed the study, compiled the data and wrote the final copy of the manuscript. Author GD performed the molecular analysis, assisted with the writing of the manuscript. Author KM monitored the variety development and field evaluation, and reviewed the manuscript. Author PM monitored the molecular genetic activities and reviewed the manuscript. Author GT assisted with molecular analysis. All authors read and approved the final manuscript.

<u>Editor(s):</u>
(1) Qiang Ge, University of Texas Southwestern Medical Center at Dallas, USA.
<u>Reviewers:</u>
(1) MSR Krishna, Department of Biotechnology, KL University, India.
(2) Anonymous, Poland.

ABSTRACT

Aims of the Study: This study was to determine the amino acid profile of two newly developed quality protein maize varieties and to develop variety-diagnostic molecular markers.
Methodology: Two new maize varieties, named MUDISHI 1 and MUDISHI 3 have been developed by breeders and farmers using the participatory breeding approach. Total protein and amino acid profiles of the two new lines were compared to their respective parental population and a locally

Corresponding author: Email: knkongolo@laurentian.ca

released genetically improved normal maize variety. Maize accessions from the DR-Congo breeding program were analyzed using ISSR primers. Variety-diagnostic markers were identified and characterized.

Results: Protein analysis data revealed that MUDISHI 1 and MUDISHI 3 are QPM varieties that are distinct from their original population, Longe 5 QPM from NARI- Unganda and DMR-ESR-W-QPM from the International Institute for Tropical Agriculture (ITTA, Ibadan), respectively. Lysine content in MUDISHI 1, and MUDISHI 3 were 3.5 g and 3.6 g of lysine / 100 g, respectively, which represents a significant increase of 20% and 23% over the genetically improved normal maize variety (Salongo 2) that is currently released. There was a significant increase of 25% of tryptophan and 33% of methionine in MUDISHI 3 compared to its parental variety while the amount of lysine was similar for the two varieties. There were 10% and 15% decrease of lysine and tryptophan, respectively in MUDISHI 1 compared to its original parent Longe 5 QPM. Genomic DNA was extracted from different maize varieties. One ISSR diagnostic-marker of 480 bp that was identified was unique to the QPM variety MUDISHI 3. This sequence was converted to a sequence characterized amplified region (SCAR) marker using a pair of designed primers. This SCAR sequence was not specific to MUDISHI 3 as it was present in all the varieties tested.

Conclusion: The newly developed varieties are typical QPM lines. The development of an ISSR diagnostic marker indicates that it is possible to develop a molecular breeding program involving QPM and normal varieties.

Keywords: Quality Protein Maize (QPM); amino acid profile; molecular markers; variety-diagnostic – marker; DR-Congo; MUDISHI 1 and MUDISHI 3.

1. INTRODUCTION

Maize is a major cereal crop for both livestock and human globally [1]. Several millions of people particularly in developing countries derive their protein and daily calorie requirements from maize [2]. It accounts for up to 15 to 56% of total daily calories in diets of people in 25 developing countries [1]. In these countries particularly African and Latin American, animal protein is scarce, expensive and unavailable to vast majority of the society [3]. Normal maize varieties are deficient in two essential amino acids, lysine and tryptophan required for human nutrition [3,4]. Lysine content in normal maize is 2% which is less than half the amount recommended for human nutrition [5]. Maize contains other amino acid, but low levels of lysine and tryptophan dilutes the contribution of other essential amino acid in maize grains [1]. Due to the poor protein quality in normal maize, there is a high prevalence of malnutrition in countries that rely solely on normal maize as sole source of daily nutrients are faced with high cases of malnutrition [1,6,7]. A genetic approach was taken to address this problem [8]. Researchers at Purdue university (USA) discovered a mutation in maize designated *opaque 2* (*O2*) [7,8]. This mutation doubled the amount of lysine and tryptophan in maize grain compared to the normal variety. They later discovered a pleiotropic effect in *O2* mutants such as soft endosperm, which made them more susceptible to pest attack and kernel damage. This was undesirable for use especially by farmers in developing countries were consumers were familiar with hard kennel normal maize varieties. Efforts to improve the poor grain quality of *O2* maize mutant, led to the development of quality protein maize (QPM) varieties by the international maize and wheat improvement center (CIMMYT) in Mexico in the 1990's [8].

Quality protein maize (QPM) germplasm was developed by the discovery of *opaque 2* (*o2*) endosperm modifier genes [1,9,10]. Two of these modifier genes have been identified. One locus mapped near centromere of chromosome 7 and the second near telomere on the long arm of chromosome 7 [11]. *O2* gene modifiers alter the phenotype of soft *O2* mutant endosperm to vitreous endosperm as well as maintain the double lysine and tryptophan content present in *O2* mutant maize verities. Through back crossing and recurrent selection, breeders at CIMMYT developed a large number of elite QPM varieties for distribution [9,12].

The plant and grain of QPM is similar in appearance and are difficult to distinguish from normal maize. Although similar phenotypically to normal maize, Nutritionally, QPM grains contains approximately 55% and 30% more tryptophan and lysine respectively compared to normal maize varieties [13]. The protein quality of QPM is 90% the nutritional value of skim milk [1].

Several countries in Africa, Latin America along with China have incorporated QPM in their Agricultural development plan [14]. However because of similar phenotypic appearance between normal and quality protein maize, a more reliable method of identifying quality protein maize has to be developed for breeding purpose.

Globally, many Asian, African, and Latin American countries are part of the network facilitated by CIMMYT, for the improvement of QPM in developing countries [1,14]. These breeding programs aimed at developing new QPM varieties that are adapted to specific environments and regional needs. Two new QPM varieties named MUDISHI 1 and MUDISHI 3 have been developed in the DR-Congo maize breeding program. These varieties are adapted to different agro-ecological regions.

The main objective of the present study was to determine the amino acid profile of MUDISHI 1 and MUDISHI 3 and to develop variety-diagnostic molecular markers that could be used to specifically track this accession in a maize breeding program.

2. MATERIALS AND METHODS

2.1 Genetic Materials

The QPM accessions used in the present study are described in Table 1. They include, GPS-5, Salongo – 2, DMR-ESR- W; DMR-ESR-W-QPM, Locale-2, AK9331-DMR-ESR-Y, MUS-1, Locale – 1, QPM LONGE-5, ECAQVE-3, ECAQVE-4, ECAQVE-6, QPM-SRSYNTH, SUSUMA, MUDISHI-1 AND MUDISHI-3. The source and year of introduction of each maize variety used in this are listed in Table 1.

2.2 Development of MUDISHI 1 and MUDISHI 3

MUDISHI 1 and MUDISHI 3 are open-pollinated quality protein varieties developed by the National Institute for Agronomic Study and Research, (INERA – DR-Congo) and Laurentian University, Sudbury, Ontario, (Canada). They were developed by breeders and farmers using the participatory breeding approach. The original variety used to develop MUDISHI 1 was QPM Longe 5 and DMR-ESR-W – QPM for MUDISHI 3. Longe 5 QPM was from NARI – Uganda while DMR –ESR-W QPM was obtained from the International Institute for Tropical Agriculture (IITA) in Ibadan, Nigeria.

MUDISHI 1 and MUDISHI 3 are the results of open pollinations of their parental lines with several QPM and normal maize varieties that were grown in the same location for few seasons. The QPM accessions include QPM Longe 5, ECAVE – 3, ECAVE-4, ECAVE-6, QPM – SR-Synth, and Susuma and the normal maize involved are DMR –ESR-W, AK9331-DMR-ESR-Y, Salongo 2, MUS 1, GPS 5, and Locale 1. The open pollinated plants were grown and progenies were selected in isolation for different agronomic characteristics for several cycles in different environments. The main selection criteria include, spike size, resistance to mildew and maize streak virus, grain yield and nutritional quality (lysine, tryptophan and other amino acid contents), and organoleptic characteristics. Plant selection and variety evaluation were performed using participatory variety selection (PVS) approach with local farmers led by breeders. The new varieties are adapted to agro-ecological conditions of Southern, Central, and Western DR-Congo.

2.3 Protein and Amino Acid Analysis

Amino acid analyses were conducted at the University of Missouri Agricultural Experiment Station Chemical Laboratories (ESCL). Total amino acid profiles were determined for the newly developed QPM varieties, MUDISHI 1 and MUDISHI 3 along with their respective parental lines (QPM Longe 5 and DMR – ES – W- QPM). One locally released and genetically improved normal maize variety (Salongo 2) was also analyzed as reference. All the analyses were conducted in triplicates. The grain amino acid concentration was evaluated using AOC standard method (Method 982.30 E (a, b, c), AOAC [15]. Crude protein was determined by combination analysis (Method 990.03, [15] using the formula crude protein = N x 6.25. ANOVA (two-way) was used to identify significant variation for each amino acid and crude protein. The least significant differences were determined to compare means.

2.4 Molecular Analysis

2.4.1 DNA extraction

Total genomic DNA from maize seedlings were extracted using the cyltrimethylammonium bromide (CTAB) protocol described by Nkongolo [16] and Nkongolo et al. [17] with some modifications. The modifications included the addition of polyvinylpyrrolidone (PVP) and beta-mercaptoethanol to the CTAB extraction buffer. DNA extracted purity was determined using spectrophotometer (Varian Cary 100 UV-VIS spectrometer).

2.4.2 Amplification of ISSR and RAPD primers

A total of 24 ISSR and 46 RAPD primers synthesized by Invitrogen were used for DNA amplification. PCR analysis was carried out following the procedure described by Mehes et al. [18] and Vaillancourt et al. [19]. Each PCR reaction was performed using a total of 25 µl volume containing 11.4 µl double distilled water, 10 mM tris-HCl pH 8.3 at 25°C taq buffer, 2 mM MgSO$_4$, and 0.5 µM of each dNTP (Applied Biosystems, Foster City, CA), 0.5 µM primer, 5ng/µl genomic DNA and 0.625 U of taq DNA polymerase (Applied Biosystems, Foster City, CA). For each primer, double distilled water was used as a negative control. Also a drop of mineral oil was added to each reaction to prevent evaporation. The samples were amplified in a thermal cycler (Perkin Elmer, Foster City, CA). The cycles performed were as follows: an initial denaturation at 95°C for 5 minutes, followed by a 2 minute incubation at 85°C at which point the taq polymerase was added; 42 cycles of 30 minutes at 95°C, 90 seconds at 55°C and 30 seconds at 72°C; a final extension for 7 minutes at 72°C was followed by subsequent incubation at 4°C.

To the PCR product, 5 µl of loading buffer was added to make a total of 30 µl. About half of this volume was loaded in a 2% agarose gel stained with 1 µl ethidium bromide. Running buffers used were 0.5X tris borate EDTA or 1X tris acetate EDTA buffer. These products were run against a 1 kb plus DNA ladder for approximately 150 minutes at 64 volts. The agarose gels were visualized and documented by using the Bio-Rad ChemiDox XRS system and analyzed with the discovery series quantity 1 D Analysis software.

2.4.3 Cloning and sequencing

A variety-diagnostic band was identified at 480 bp in maize variety MUDISHI 3 by amplifying maize genomic DNA with ISSR primer HB 15 (Fig. 1). This band was cloned and sequenced as described by Vaillancourt et al. [19] with the following modifications: Unique diagnostic band was run in low melt 3% agarose gel. It was excised and gel plugs were dissolved with 1X tris-EDTA (TE) buffer. DNA extractions from gel plugs were achieved through several chloroform and phenol DNA extractions described by

Vaillancourt et al. [19]. After sequencing, a primer pair was designed and synthesized to target the insert region by using Life technology software (OligoPerfect™ Designer). The primer pair was used to amplify normal and quality protein maize DNA to verify the specificity of the SCAR markers.

3. RESULTS

3.1 Agronomic Characteristics

MUDISHI 1 and MUDISHI 3, two new maize varieties with white grain color were released in 2012 and they are adapted to all the maize growing regions in western, central, and southern DR-Congo. Days to maturity for these varieties in these areas average 115 and 100, for MUDISHI 1 and MUDISHI 3, respectively. The yield in farmer's field without fertilizers is 1 T to 1.5 T for MUDISHI 1 and 0.8 to 1 T/ ha for MUDISHI 3. This yield can reach up to 6 T / ha for MUDISHI 1 and 3 T to 4 T for MUDISHI 3 at research field station under mineral fertilization. MUDISHI 1 is susceptible to down mildew and maize streak virus. MUDIDHI 3 is highly resistant to downy mildew and to lodging and resistant to maize streak virus.

3.2 Amino Acid Profiles

The overall amino acid composition of the maize varieties and the levels of statistical significance obtained from analysis of variance are shown in Table 2. Lysine content in MUDISHI 1 and MUDISHI 3 were 3.6 g and 3.5 g of lysine / 100 g, respectively, which represents a significant increase of 23% and 20%, over the genetically improved normal maize variety (Salongo 2) that is currently released. There was a significant increase of 25% of tryptophan in MUDISHI 3 compared to its parental variety (DRM-ESR-W–QPM) while the amount of lysine was similar for the two varieties. But, there were 10% and 15% decrease of lysine and tryptophan, respectively in MUDISHI 1 compared to its original parent Longe 5 QPM.

The other potentially limiting amino acids are threonine, isoleucine and methionine. Threonine and isoleucine levels were relatively similar in MUDISHI 1, MUDISHI 3, DMR-ESR-W QPM, Longe 5 QPM, and Salongo 2. A significant increase of 20% and 33% of methionine in MUSHISHI 3 over Salongo 2 and the original parent (DMR-ESR-W-QPM), respectively was observed. Likewise there were 34% more methionine in MUDISHI 1 compared to its parental line, Longe 5 QPM. The levels of leucine and glutamic acid were the same in MUDISHI 3 compared to its original parental DMR-ESR-W-QPM. But the values for these elements were slightly lower in MUDISHI 1, Longe 5 QPM, and Salongo 2. The crude protein content were 9.1%, 8.9%, 9.5%, 8.8%, and 9.6% in MUDISHI 1, MUDISHI 3, DMR-ES-W-QPM, Longe 5 QPM, and Salongo 2, respectively (Table 2).

Table 1. List of maize accessions used in this study

Varieties	Origin	Year of introduction	Type
GPS-5	INEAC-Gandajika-DR-Congo	-	NORMAL
SALONGO-2**	INERA- Gandajika-DR-Congo	1976	NORMAL
DMR-ESR-W**	IITA-Ibadan	1994	NORMAL
LOCALE-2	Farmers- Gandajika-DR-Congo	-	NORMAL
AK9331-DMR-ESR-Y	IITA-Ibadan	1994	NORMAL
MUS-1**	INERA	1996	NORMAL
LOCALE-1	Farmers- Gandajika-DR-Congo	-	NORMAL
DMR-ESR-W-QPM	IITA –Ibadan	1994	QPM
QPM-LONGE 5***	NARI-Uganda	2008	QPM
ECAQVE-3	CIMMYT-Kenya	2008	QPM
ECAQVE-4	CIMMYT-Kenya	2008	QPM
ECAQVE-6	CIMMYT-Kenya	2008	QPM
QPM-SR-SYNTH***	CIMMYT-Kenya	2008	QPM
SUSUMA	CIMMYT-Kenya	2008	QPM
MUDISHI 1	INERA-DR-Congo/Laurentian University	2012	QPM
MUDISHI 3	INERA-DR-Congo/Laurentian University	2012	QPM

**Selected improved normal maize varieties, ***Selected Elite Quality protein maize varieties, CIMMYT: International Maize and Wheat Improvement Center, INERA: National Institute of Agronomic Research and Studies (DR-Congo), INEAC: National Institute of Agronomic Studies (Belgium, Congo), IITA: International Institute of Tropical Agriculture and NARI: Namulonge Agriculture Research Institute

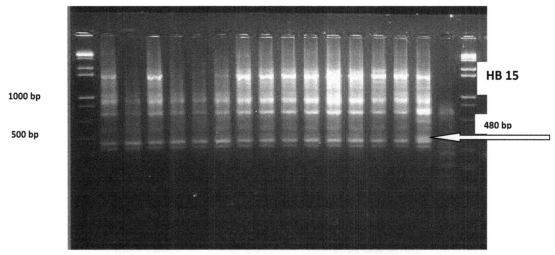

Fig. 1. ISSR amplification of maize DNA samples with the HB 15 primer-generating a 480 diagnostic marker

Lanes 0 and 17 contain 1-Kb plus ladder; Lanes 1 to 15 contain GPS-5, SALONGO-2**, ECAQVE-6, AK9331-DMRESR-Y, MUS-1**, LOCALE-1, QPM-LONGE5***, ECAQVE-3, ECAQVE-4, DMR-ESR-W**, LOCALE-2, QPM-SRSYNTH***, SUSUMA, MUDISHI 1 and MUDISHI 3; Lane 16, blank. Arrow on the right indicates variety-diagnostic marker at 480 bp

Table 2. Total protein and essential amino acid content in quality protein maize (QPM) and normal maize varieties from the DR-Congo breeding program

Essential AA w/w (%)*	Corn varieties					LSD
	Mudishi 1	Longe 5 qpm	Mudishi 3	Dmr-esr-w-qpm	Salongo 2	
Taurine	0.11 (1.2)	0.12 (1.4) (1.36)	0.11 (1.2)	0.12 (1.3)	0.03 (0.3)	0.05
Hydroxyproline	0.04 (0.44)	0.04 (0.5)	0.02 (0.3)	0.04 (0.4)	0.03 (0.3)	0.05
Aspartic acid	0.63 (6.9)	0.63 (7.2)	0.56 (6.3)	0.64 (6.8)	0.60 (6.3)	0.58
Threonine	0.33 (3.6)	0.32 (3.7)	0.31 (3.5)	0.35 (3.7)	0.34 (3.6)	0.30
Serine	0.40 (4.4)	0.39 (4. 5)	0.41 (4.6)	0.42 (4.4)	0.44 (4.6)	0.44
Glutamic acid	1.62 (17.8)	1.51 (17.2)	1.63 (18.2)	1.74 (18.4)	1.89 (19.7)	1.00
Proline	0.85 (9.3)	0.87 (9.9)	0.82 (9.4)	0.86 (9.1)	0.86 (9.0)	0.70
Lanthionine	0.00 (0.0)	0.00 (0.0)	0.00 (0.0)	0.00 (0.0)	0.00 (0.0)	–
Glycine	0.38 (4.2)	0.39 (4.5)	0.38 (4.3)	0.40 (4.2)	0.35 (3.7)	0.45
Alanine	0.63 (6.9)	0.57 (6.5)	0.65 (7.3)	0.68 (7.2)	0.75 (7.8)	0.71
Cysteine	0.24 (2.6)	0.23 (2.6)	0.22 (2.5)	0.22 (2.3)	0.20 (2.1)	0.40
Valine	0.46 (5.1)	0.47 (3.4)	0.44 (5.0)	0.48 (5.1)	0.47 (4.9)	0.45
Methionine	0.21 (2.3)	0.15 (1.7)	0.20 (2.3)	0.16 (1.7)	0.18 (1.9)	0.10
Isoleucine	0.32 (3.5)	0.30 (3.4)	0.31 (3.5)	0.34 (3.6)	0.36 (3.8)	0.40
Leucine	1.03 (11.3)	0.90 (10.3)	1.05 (11.9)	1.11 (11.7)	1.31 (13.7)	0.81
Tyrosine	0.20 (2.2)	0.20 (2.3)	0.22 (2.5)	0.23 (2.4)	0.26 (2.7)	0.15
Phenylalanine	0.43 (4.7)	0.39 (4.5)	0.43 (4.9)	0.46 (4.9)	0.50 (5.2)	0.59
Hydroxylysine	0.02 (0.2)	0.02 (0.2)	0.02 (0.2)	0.02 (0.2)	0.02 (0.2)	0.17
Ornithine	0.01 (0.0)	0.01(0.1)	0.01 (0.1)	0.01 (0.1)	0.01 (0.1)	0.00
Lysine	0.33 (3.6)	0.35 (4.0)	0.29 (3.5)	0.33 (3.5)	0.28 (2.9)	0.29
Histidine	0.33 (3.6)	0.34 (3.9)	0.29 (3.3)	0.33 (3.5)	0.27 (2.8)	0.35
Arginine	0.46 (5.1)	0.49 (5.6)	0.42 (4.7)	0.46 (4.9)	0.39 (4.1)	0.60
Tryptophan	0.07 (0.8)	0.08 (0.9)	0.07 (0.8)	0.06 (0.6)	0.05 (0.5)	0.05
Total	9.10	8.77	8.86	9.46	9.59	
Crude protein	9.83	9.45	9.55	10.25	9.89	

*The values are expressed in w/w = grams per 100 grams of sample. The number is parentheses represent the percent (%) of individual amino acid in the crude protein. AA = Amino Acid. Longe 5 QPM and DMR-ESR-W-QPM are parental lines for MUDISHI 1 and MUDISHI 3, respectivel

Overall, the total basic acids, which include lysine, arginine, and histidine constituent 12.3% and 11.2 % of the total amino acids for MUDISHI 1 and MUDISHI 3, respectively. These values were 11.9% and 13.5% for DMR-ESR-W QPM and Longe 5-QPM varieties, respectively. They were lower (9.8%) in normal maize Salongo 2. In general, the total basic acids are considerably lower than the acidic amino acids (aspartic acid and glutamic acid), which represent around 25% of the total amino acid residue for both MUDISHI 1 and MUDISHI 3. The acidic amino acid levels were 25.2%, 24.4%, and 26%, for DMR-ESR-W QPM, Longe 5 QPM, and Salongo 2, respectively.

3.3 Molecular Analysis

3.3.1 ISSR and RAPD primer analysis

The ISSR and RAPD primers used are described in Tables 3 and 4. Out of the 70 primers screened, 24 ISSR and 46 RAPD primers generated amplified products. One of the 17 ISSR primers HB 15 generated a diagnostic marker at 480 bp for MUDISHI 3 (Fig. 1) that was selected for further analysis. The other primers either generated poor amplification or did not produce unique band for QPM variety identification.

3.3.2 Identification of variety-diagnostic markers

The variety-diagnostic marker of 480 bp size that was diagnostic for MUDISHI 3 DNA sample was cloned and sequenced. The consensus sequence described in Fig. 2 has been registered in the National Center for Biotechnology Information (NCBI) Genbank in Bethesda (Maryland, USA) under the accession number KM360096. BLAST search results reveal 88% matching with a deoxyribonuclease from *Pantoea vagans* C9-1. The sequence showed also 73% and 78% similarity with deoxyribonuclease from *Yersinia enterocolica* LC20 and *Serratioa marcescens*, respectively.

A primer pair targeting the insert was designed to produce a sequence characterized amplified region (SCAR) marker (Table 5). The primer pair amplified the targeted band in all the DNA samples from the QPM and normal maize varieties screeed. The identified marker is therefore diagnostic for MUDISHI 3 but not variety-specific when converted to a SCAR marker (Fig. 3).

4. DISCUSSION

4.1 General Characteristics

The two newly developed QPM varieties (MUDISHI 1 and MUDISHI 3) are the results of open pollinations of their parental lines with several QPM and normal maize varieties that were grown in the same location for few seasons. Based on amino acid profile, they have all the characteristics of quality protein maize varieties. The agronomic evaluation of these two lines revealed that they are adapted to several agro-ecological conditions in the DR-Congo and they are resistant to local maize diseases. In facts, hundreds of different open-pollinated varieties were developed by farmers in the United States during the 19th and early 20th centuries using the same approach described for MUDISHI 1 and MUDISHI 3 development. The method consisted in selecting for different plant characteristics in different environments. Some of the more famous of these American corn varieties were Krug, Lancaster Sure Crop, Leaming, Midland, and Reid [20].

4.2 Amino Acid Analysis

In several reports, lysine levels have been associated with tryptophan levels. The data reported in the present study are in accord with the lysine values reported by Mbuya et al. [2] and Kniep and Mason [21]. The significant increase of basic totally charged and hydrophilic amino acids in MUDISHI 1 and MUDISHI 3 compared to normal maize is consistent with other QPM varieties analyzed and suggest an increase in nonzein protein and hydrophobicity in QPM [2,22]. The concentration of lysine in the maize endosperm has been shown to be highly correlated with the content of a single nonzein protein called the protein synthesis factor EF-1α [23,24]. All this data indicate that MUDISHI 1 and MUDISHI 3 are QPM maize varieties.

Even though lysine content in MUDISHI 1 and MUDISHI 3 proteins were significantly higher than normal maize, it is still below the recommended FAO/WHO reference lysine standard value of 58 mg/g of dietary protein for a 2 – 5 year child [5]. Specifically, MUDISHI 1 and MUDISHI 3 supply 62% and 60% of human lysine requirements, respectively compared to 48% for Salongo 2.

Although normal maize is not deficient in isoleucine or threonine, the presence of large amount of leucine in human diet can cause both amino acid imbalances and interference of isoleucine absorption. The ratio of leucine / isoleucine found in MUDISHI 1 and MUDISHI 3 were 3.2 and 3.3, respectively.   This indicates that these two new varieties provide proteins with a better EAA balance compared to normal maize. This is consistent with Huang et al. [25] stating that a pleiotropic increase in non-zein proteins is contributing to an improved amino acid balance.

4.3 Molecular Analysis

To date, most studies on QPM in relation to molecular markers have been restricted to identifying genetic distance and diversity among QPM and normal maize [26,27]. Nkongolo et al. [26], studied genetic diversity among QPM and normal maize accessions from Africa, and found low genetic distance and diversity between accessions which have also been reported in other studies [28].

In the present study, only 50% of the RAPD primers generated PCR amplification products compared to 74% of ISSR primers that amplified DNA samples. This could be due to more sensitivity of RAPD primers to PCR contaminants compared to ISSR primers [29].

We have identified an ISSR marker that is diagnostic for MUDISHI 3 in a breeding program. This ISSR marker was not specific once converted to a SCAR marker. This suggests that it is present in other varieties but in a low copy number. The marker will be useful as a diagnostic tool to track MUDISHI 3 genome in progenies derived from crosses involving this variety.

There exists couple of reasons that could explain why diagnostic marker such as the 480 bp sequence in MUDISHI 3 once converted to a SCAR marker is not specific. The appearance of a band in one species or a variety and its absence in another could be the result of competition among DNA fragments during amplification [30]. Amplified products which are complementary to each other are stabilized by internal base pairing that could prevent amplification by out-competing the binding of random primers [30,31]; this is the most serious problem that leads to the incorrect interpretation of results. Formation of secondary structures including hair pin, by DNA fragments. Since the SCAR system is a more sensitive technique, the above problems will not interfere with the amplification of sequences even if they are present in a low copy number.

Table 3. The 3' anchored nucleotide sequences of ISSR primers used to screen DNA samples from all the maize varieties

ISSR Primer	Nucleotide sequence (5'-3')	Fragment size range (bp)
809	(AG)8G	250-850
818	(CA)8G	250-100
823	(TC)8C	-
827	(AC)8G	100-650
829	(TG)8C	300-850
834	(AG)8YT	300-1650
835	(AG)8YC	200-1000
841	GAA GGA GAG AGA GAG AYC	300-1000
844	(CT)8RC	300-850
849	(GT)8YA	200-1000
873	(GACA)4	-
879	(CTTCA)3	100-2000
17898B	(CA)6GT	200-1650
ECHT 3	(AAC)3GC	-
ECHT 4	(AAG)3GC	-
HB 15	(GTG)3GC	200-1000
SC ISSR 3	(GAC)4G	400-5000
SC ISSR 4	(CGT)4C	-
SC ISSR 5	(ACG)4AC	300-5000
SC ISSR 6	(TTG)5CB	-
SC ISSR 7	(AGG)5GY	-
SC ISSR 8	(AGAT)5GY	-
SC ISSR 9	(GATC)3GC	300-2000
SC ISSR 10	(CTT)5(CCT)6CT	-

Table 4. Nucleotide sequences of RAPD primers used to screen DNA samples from all the maize varieties

RAPD primers	Nucleotide sequence	Fragment size (bp)
C10	TGTCTGGGTG	400-2000
D8	GTGTGCCCCA	-
E12	TTATCGCCCC	500-5000
F10	GGAAGCTTGG	500-2000
Grasse 1	CCGCCCAAAC	-
Grasse 2	GTGGTCCGCA	200-1650
Grasse 3	GTGGCCGCGC	-
Grasse 4	GAGGCGCTGC	400-1650
Grasse 5	CGCCCCCAGT	-
Grasse 7	CACGGCGAGT	-
Grasse 9	GTGATCGCAG	1000-2000
OPA 1	CAGGCCCTTC	400-1000
OPA 3	AGTCAGCCAC	400-1000
OPA 8	GTGACGTAGG	-
OPA 11	CAATCGCCGT	400-5000
OPA 12	TCG GCGATAG	-
OPA 14	TCTGTGCTGG	-
OPA 15	TTCCGAACCC	-
OPA 16	AGCCAGCGAA	-
OPA 17	GACCGCTTGT	-
OPA 18	AGGTGACCGT	400-2000
OPA 19	CAAACGTCGG	-
OPA 20	GTTGCGATCC	300-2000
OPB 1	GTTTCGCTCC	500-1650
OPB 2	TGATCCCTGG	100-1650
OPB 3	CATCCCCCT G	-
OPB 4	GGACTGGAGT	200-1000
OPB 6	TGCTCTGCCC	400-1650
OPB 7	GGTGACGCAG	300-1000
OPC 10	TGTCTGGGTG	-
OPE 9	CTTCACCCGA	-
OPT 17	CCAACGTCGT	300-5000
OPX 4	CCGCTACCGA	400-5000
OPY 9	AGCAGCGCAC	300-2000
PINUS 23	CCCGCCTTCC	-
PINUS 146	ATGTGTTGCG	-
UBC 48	TTAACGGGGA	-
UBC 78	GAGCACTAGC	-
UBC 186	GTGCGTCGCT	300-1650
UBC 214	CATGTGCTTG	-
UBC 270	TGCGCGCGGG	300-850
UBC 402	CCCGCCGTTG	-
UBC 486	CCA GCATCAG	-
UBC 494	TGATGCTGTC	-
UBC 551	GGAAGTCCAC	400-1000
UBC 561	CATAACGACC	-

Table 5. Nucleotide sequences of the designed primers targeting variety-diagnostic ISSR marker

Primer name	Bases	Sequence (5'-3')
Maize 480 6F	20	TCA TTG TTC ACA CCC GTG AT
Maize 480 6R	20	GCC AGC GTT TCT AAA TCC AC

GTGGTGGTGG CTCTGGGGGA AACCGgGCTG GATTATCACT ATCAGCCAGA AACAAAAGAT
CAGCAGCAGC GCTCGTTCCT GGAACATATC CGTACCGGTA TTGCGCTGAA CAAACCGA**TC**
ATTGTTCACA CCCGTGATGC CCGCGAAGAT ACCCTGACGA TTCTGCGTGA AGAGCAGGTT
GAACGTTGCG GCGGCGTGCT GCACTGCTTC ACTGAGGATC AGCCCACCGC AGCAAAACTG
CTGGATATGG GCTTTTACAT CTCTTTTTCC GGCATCGTCA CATTCCGCAA TGCCGAGCAG
TTACGTGAAG CCGCACGCTA TGTGCCGCTG GATCGGATGC TGGTGGAAAC GGATTCGCCT
TATCTGGCAC CGGTGCCTTT CCGTGGTAAA GAGAATCAGC CCGCTTATAC GCGCGATGTT
GCCGAATATC TGGCTATCCT GAAAGGG**GTG GATTTAGAAA CGCTGGC**AGCCACCACCAC

Fig. 2. Consensus sequence of variety-diagnostic ISSR marker of 480 bp from MUDISHI 3 generated by the ISSR primer HB 15; Underlined region indicates the ISSR HB 15 primer consensus sequence. Bolded region indicates the SCAR marker region 349 bp

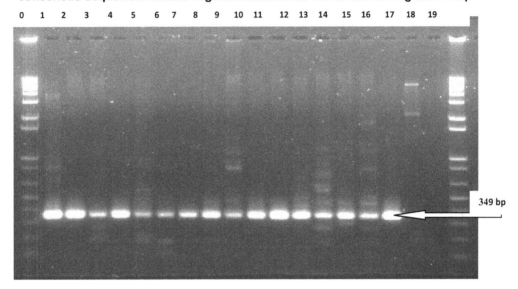

Fig. 3. SCAR bands generated using a designed pair of primers targeting the MUDISHI 3 diagnostic marker. Lane 0 and 19 contain 1-Kb plus ladder; lane 1, Recombinant plasmid with 480 bp marker; lane 2, MUDISHI 3; lane 3-16, maize varieties 15 contain GPS-5, SALONGO-2, ECAQVE-6, AK9331-DMR-ESR-Y, MUS-1, LOCALE-1, QPM-LONGE-5, ECAQVE-3, ECAQVE-4, DMR-ESR-W**, LOCALE-2, QPM-SRSYNTH***, SUSUMA, MUDISHI 1 and MUDISHI 3 1-15 (Table 1); lane 17, background plasmid (no insert); lane 18, blank. The arrows indicate the SCAR marker that is present in all the DNA samples**

The need for molecular identification of maize variety by using variety-specific markers such as SCAR has increased recently in maize breeding programs. This is partly due to the development of QPM hybrid adapted to local regions that require the application of molecular tools that are more reliable than other methods. Another reason would be due to the open pollinated nature of maize plant that leads to contamination of elite QPM maize lines if not properly handled [1]. Protein analysis which indicates correctly the presence of improved protein is very expensive and not recommended for most breeding programs [32]. The application of diagnostic markers could be quite expensive and time consuming. It can be only used if a SCAR marker cannot be developed. Hence, for molecular breeding, SCAR marker remains the most effective method to track QPM genome in maize hybrid background [19].

The low level of genetic variation among accessions that made it difficult to develop more diagnostic markers could be explained by the effect of bottle neck among African maize accessions [26]. More studies would need to be carried out to develop SCAR markers for QPM by possibly using more unanchored ISSR primers.

5. CONCLUSION

The premise of this study was to determine the amino acid profile of a new QPM and to develop variety diagnostic/ specific molecular markers for quality protein and normal maize using ISSR and

RAPD primers. Protein analysis data revealed that MUDISHI 1 and MUDISHI are QPM varieties that are distinct from their original populations (Longe 5 QPM and DMR-ESR-W-QPM). One primer revealed a diagnostic marker for QPM MUDISHI 3. The remaining primers showed relatively good amplification and a high level of polymorphism. Primers flanking the diagnostic marker sequence were developed. However the SCAR marker amplified was present in all the maize accessions analyzed. Although the diagnostic marker that we have developed is useful in tracking MUDISHI 3 genome in progenies, further analysis of several other ISSR and RAPD primers is required to achieve the main goal of developing variety-specific markers in the targeted breeding program.

ACKNOWLEDGEMENTS

This research was conducted through a partnership between Laurentian University (Ontario, Canada), INERA-DR-Congo, University of Kinshasa (DR-Congo), and Caritas Congo. The authors are grateful to the Canadian International Development agency (CIDA) for financial support and INERA – Gandajika for providing the logistics for the development of MUDISHI 1 and MUDISHI 3. We thank the African Development Bank (ADB) for providing additional financial support through the Forum for Agricultural Research in Africa (FARA) and the Association for Strengthening Agricultural Research in Eastern and Central Africa (ASARECA) for further nutritional evaluation of MUDISHI 3 before its official release. Thanks to Mrs. Ramya Narendrula for assistance with manuscript preparation.

COMPETING INTERESTS

Authors have declared that no competing interests exist.

REFERENCES

1. Prasanna BM, Vasal SK, Kassahun B, Singh NN. Quality protein maize. Curr Sci. 2001;81(10):1308-1318.
2. Mbuya K, Nkongolo KK, Kalonji-Mbuyi A. Nutritional analysis of quality protein maize varieties selected for Agronomic characteristics in a breeding program. Int J Plant Breed Genet 2011;5(4):317-327.
3. FAO Agrostat, Food Balance Sheets, FAO, Rome, Italy; 1992.
4. Gaziola SA, Alessi ES, Guimaraes PEO, Damerval C, Azevedo RA. Quality protein maize (QPM): A biochemical study of enzymes involved in lysine metabolism. J Agric Food Chem. 1999;47:1268-1275.
5. FAO/WHO. Protein Quality Evaluation. Expert Consultation, Report Series 51, Food and Agriculture Organization/ World Health Organization Nutrition Meetings, Rome, Italy; 1991.
6. Akalu G, Taffesse S, Gunaratna NS, De Groote H. The effectiveness of quality protein maize in improving nutritional status of young children in Ethiopian Highlands. Food Nutr Bull. 2010;31(3):418-430.
7. Nuss ET, Tanumihardjo ST. Closing the protein inadequacy gap in vulnerable populations. Adv Nutr. 2011;2:217-224.
8. Bjarnason M, Vasal SK. Breeding for quality protein maize. Plant Breed Rev. 1992;9:181-216.
9. Greetha KB, Lending CR, Lopes MA, Wallace JC, Larkins BA. Opaque-2 modifiers increase y-zein synthesis and alter its special distribution in maize endosperm. Plant Cell. 1991;3:1207-1219
10. Bostwick E, Larkins BA. Opaque-15, a maize mutation with properties of a defective opaque-2 modifier (y zein/Quality Protein Maize/storage proteins). Proc Natl Acad Sci USA. 1995;92:1931-1935.
11. Dannenhoffer JM, Bostwick DE, Or E, Larkins BA. Opaque-15, a maize mutation with properties of a defective opaque-2 modifier (y-zein/Quality Protein Maize/ storage proteins). Proc Natl Acad Sci. USA. 1993;92:1931-1935.
12. Vasal SK, Villegas E, Bjarnason M, Gelaw B, Goertz P. Genetic Modifiers and Breeding Strategies in Developing Hard Endosperm opaque-2 Materials, pp. in Improvement of Quality Traits of Maize for Grain and Silage Use, edited by Pollmer W. G., Phipps R. H., MartinusNijhoff, London. 1980;37-73.
13. Paes MC, Bicudo MH. Proceedings of the international symposium on quality protein maize (eds Larkins, B. A and Mertz, E. T.), EMBARAPA/CNPMS, Sete Lagaos, Brazil. 1995;65-78.
14. Mbuya K, Nkongolo KK, Kalonji-Mbuyi A, Kizungu R. Participatory, selection and characterization of quality protein maize (QPM) varieties in Savanna agro ecological region of DR-Congo. J. Plant Breed Crop Sci. 2010;2(11):325-332.

15. AOAC. Official Methods of Analysis. 18th Edn., Association of Official Analytical Chemists, Gaithersburgs, MD; 2006.

16. Nkongolo KK. RAPD and cytological analysis of *Picea* spp. from different provenances: Genomic relationships among taxa. Hereditas. 1999;130:137-144.

17. Nkongolo KK, Gervais S, Zhou Y. Comparative analysis of inter simple sequence repeats and simple sequence repeats markers: genetic analysis of *Deschampsia cespitosa* populations growing in metal contaminated regions in Canada. Amer J Biochem Biotechnol. 2014;10(1):69-80.

18. Mehes M, Nkongolo KK, Michael P. Genetic variation in Pinusstrobus and P. Monticola populations from Canada: development of genome- specific markers. Plant Syt Evol. 2007;267:47-63.

19. Vaillancourt A, Nkongolo KK, Michael M. Identification, characterization, and chromosome locations of rye and wheat specific ISSR and SCAR markers useful for breeding purposes. Euphytica. 2008;159:297-306.

20. Russell WK, Sandall L. Corn breeding: types of cultivars. J Nat Resour Life Sci Educ. 2006;35(1):242.

21. Kniep KR, Mason SC. Lysine and protein content of normal and opaque-2 maize grain as influenced by irrigation and nitrogen. Crop Sci. 1991;31:177-181.

22. Zarkadas CG, Hamilton RI, Yu ZR, Choi VK, Khanizadeh S, Rose NGW, Pattison PL. Assessment of the protein quality of 15 new Northern adapted cultivars of quality protein maize using amino acid analysis. J Agric Food Chem. 2000;48:5351-5361.

23. Habben JE, Moro GL, Hunter BG, Hamaker BR, Larkins BA. Elongation factor 1 α concentrations is highly correlated with the lysine content of maize endosperm. Proc Natl. Acad Sci. USA. 1995;92:8640-8644.

24. Moro GL, Habben JE, Hamaker BR, Larkins BA. Characterization of the variability in lysine content for normal and opaque-2 maize endosperm. Crop Sci. 1996;36:1651-1659.

25. Huang S, Frizzi, A, Florida CA, Kruger DE, Luethy MH. High lysine and high tryptophan transgenic maize resulting from the reduction of both 19- and 22-kD alpha-zeins. Plant Mol Biol. 2006;61:525-535.

26. Nkongolo KK, Mbuya K, Mehes-Smith M, Kalonji-Mbuya A. Molecular characterization of quality protein maize (QPM) and normal maize varieties from the DR-Congo breeding program. Afr J Biotechnol. 2011;10(65):14293-14301.

27. Mbuya K, Nkongolo KK, Narendrula R, Kalonji-Mbuyi A, Kizungu RV. Development of quality protein maize (QPM) inbred lines and genetic diversity assessed with ISSR markers in maize breeding program. Am J Exp Agric. 2012;2(4):626-640.

28. Semagn K, Bjornstad A, Ndiondjop MN. An overview of molecular markers methods for plants. Afr J Biotechnol. 2006;5(25):2540-2568.

29. Liu Y, Li S, Zhu T-h, Baolin S. Specific markers for detection of bacterial canker of kiwifruit in Sichuan, China. Afr J Microbiol Res. 2012;6(49):7512-7519.

30. Nkongolo KK, Deverno L, Michael P. Genetic validation and characterization of RAPD markers differentiating black and red spruces: molecular certification of spruce trees and hybrids. Plant Syst Evol. 2003;236:151-163.

31. McGrath A, Higgins DG, McCarthy TV. Sequence analysis of DNA randomly amplified from the Saccharomyces cerevisiae genome. Mol Cell Probes. 1998;12:397-405.

32. Krivanek AF, DeGroote H, Gunaratna NS, Diallo AO, Friesen D. Breeding and disseminating quality protein maize (QPM) for Africa. Afr J Biotechnol. 2007;6(4):312-324.

8

Low Cost Micropropagation of Local Varieties of Taro (*Colocasia esculenta* spp.)

Alex Ngetich[1*], Steven Runo[1], Omwoyo Ombori[2], Michael Ngugi[3], Fanuel Kawaka[4], Arusei Perpetua[5] and Gitonga Nkanata[3]

[1]*Department of Biochemistry and Biotechnology, Kenyatta University, P.O.Box 43844-00100, Nairobi, Kenya.*
[2]*Department of Plant Sciences, Kenyatta University, P.O.Box 43844-00100, Nairobi, Kenya.*
[3]*Department of Agriculture, Meru University College of Sciences and Technology, P.O.Box 972- 60200, Meru, Kenya.*
[4]*Department of Pure and Applied Sciences, Technical University of Mombasa, P.O.Box 90420, 80100, Mombasa, Kenya.*
[5]*Department of Botany, Moi University, P.O.Box 3900, 30100, Edoret, Kenya.*

Authors' contributions

All the authors collaborated in carrying out this work. Authors SR, OO and GN designed the study and supervised laboratory and greenhouse experiments. Authors AN and MN carried out the laboratory and greenhouse experiments and wrote the first draft. Authors FK and AP performed data analyses of the study and drafted the final manuscript. All authors read and approved the final manuscript.

Editor(s):
(1) Giovanni DalCorso, Department of Biotechnology, University of Verona, Italy.
Reviewers:
(1) Anonymous, Malaysia.
(2) Anonymous, China.

ABSTRACT

Aims: This study was conducted to evaluate low cost protocol for the micropropagation of three varieties of taro (Dasheen, Eddoe and wild) from eastern Kenya.
Study Design: The plants were grown in polythene bags arranged in a completely randomized block design (CRBD) replicated nine times.
Place and Duration of Study: Department of plant sciences Kenyatta University in plant and tissue culture laboratory, between June 2010 and December 2011.
Methodology: The three media types tested were Omex foliar feed (LCM1), Stanes micronutrients (LCM2) and micro food (LCM3) as substitute for Murashige and Skoog (MS) media.

Corresponding author: Email: akngetich@gmail.com

Results: The results showed significant differences (p<0.05) in the shoot generation for Eddoe and wild varieties in LCM1 and LCM2 respectively compared to LCM3 and MS. Plants grown in MS media and LCM3 had the longest height compared to LCM1 and LCM2. Naphthalene Acetic Acid (NAA) and Citishooter did not show any significant differences on the number of roots. All the regenerated plants in this study were similar in morphology and vigour. Media cost was reduced by 94.7% (LCM1) and 96% for both LCM2 and LCM3.

Conclusion: This study indicates the potential of low cost media as a substitute for conventional micro propagation.

Keywords: Citishooter; conventional; micronutrients; micro propagation.

1. INTRODUCTION

Taro is an important staple food crop grown throughout many Pacific Island countries, Asia, the Caribbean and many parts of Africa, for its fleshy corms and nutritious leaves. In addition to contributing to sustained food security and export earnings [1]. The crop is one of the principal root crops that have shown great potential in generating income within the rural communities [2]. In Kenya, taro is a neglected crop grown primarily by farmers in marginal areas and its large-scale cultivation is constrained by the lack of high quality seed and the low productivity and profit. The plant is very susceptible to a wide range of pests, pathogens and diseases [3], such as *Pythium* rot, dasheen mosaic virus and nematode diseases [3]. In attempt to address these challenges, routine methods such as Tissue culture has been suggested however the cost of production is high for most of the countries in the sub-Saharan Africa including Kenya. The cost of the micro-propagules production has precluded the adoption of the technology for large scale micro-propagation [4]. Media cost including chemicals and energy account for 30–35% of the cost of micro-propagation of plants [5,6]. Many studies have reported that the production cost of tissue-cultured plants can be reduced by 50-90% using low cost media ingredients and containers [7]. The conventional method of taro cultivation is through vegetative propagation. The division of taro offshoots is not always suitable for this cultivation due to the weakness and susceptibility to pathological agents. However, there is limited availability of clean planting materials. The aim of this study was to provide a reliable, low cost and high quality taro planting materials.

2. MATERIALS AND METHODS

2.1 Study Site and Sample Collection

Plantable setts of Dasheen, Eddoe and Purple/Wild varieties of taro visibly free from diseases were collected from Meru central region, Eastern Kenya. Meru district receives an average annual rainfall ranging between 380 mm -2500 mm. The plants were transported to Kenyatta University Plant Sciences net house. The plants were grown a completely randomized block design (CRBD) in polythene bags containing soil enriched with diammonium phosphate (DAP) fertilizer consisting of four treatments with nine replications. The plants heights, number of shoots and roots were monitored.

2.2 *In vitro* Propagation of Taro

Plants were washed with tap water and outer leaves removed until inner cleaner section appears with 5cm of shoot and corm of 2 cm. Plants were then surface sterilized in 2.31% NaOCl (60% commercial Jik) containing a few drops of Tween 20 for 45 minutes under a laminar flow with frequent agitation. Outer leaves were separated from the dome in a circular fashion using a sterile surgical knife. The explants were then transferred to 90% of ethanol for 1.5 min after which it was further sterilized in 70% ethanol for 12.5 min. The final trimming was done until the meristem domes of about 1 cm^2 were obtained which were rinsed 4-5 times with autoclaved double distilled water.

2.3 Low Cost Media Formulation for Taro

Three low cost substitutes for MS salts were tested. Omex foliar feed 24-24-18 + trace elements (LCM1) from Murphy Chemicals (E.A) Limited–a complete substitute for MS salts since it contains both macronutrients and micronutrients. The second treatment (LCM2) consisted of Stanes micronutrients from Osho Chemicals Limited while macronutrients came from low cost alternatives in the market that are used as fertilizers available in agrovet shops. The last treatment consisted of microfood ® horticulture from Osho Chemicals as source of micronutrients while macronutrients came from low cost alternatives in the market (LCM3) as shown in Table 1. Conductivity and pH of the fully substituted media was measured and adjusted to 5.7-5.8 using KOH and HCl then autoclaved at 15 psi and 121°C.

2.4 Culture Conditions

For all treatments that is conventional, LCM1 LCM2 and LCM3 55 ml of medium was dispensed in 5.5x10 cm glass flasks (200 ml flask); one explant (approx. 1 cm long) was cultured on each flask. Sterile meristem explants were subsequently cultured on the MS basal medium supplemented with 8 mg/l Benzo Amino Purine (BAP) and 30g/l sucrose. The cultures were then taken to growth room which growth conditions were: 25±2°C, 18 h (day)/6 h (night) photoperiod with light source provided by irradiation intensity of 40~44 µmol m^{-2} s^{-1}. Shooting was induced by transferring plants to new media after seven days and the treatments replicated four times. Multiple shoots was recorded after 4 weeks. Regenerated shoots were then transferred onto MS medium supplemented with 0.25 mg/l naphthalene acetic acid (NAA) and 30 g/l sucrose for induction of roots. Growth parameters such as number of shoots number and size, roots number and plant height, were recorded at 15 days interval for 2 months after raising the cultures.

2.5 Hardening, Acclimatization and Morphological Characterization

Regenerated plants were removed from individual glass jars. Roots were then rinsed with warm water to remove excess media and planted in small green pots containing vermiculite: Sand in the same ratio then transferred to greenhouse under 70% shade. They were covered with polythene paper to maintain humid conditions and reduce excess water loss for 2-3 weeks. They were then transplanted into medium sized black polythene pots containing soil with 5 g/kg DAP fertilizer and monitored for a month with

daily watering every morning. Subsequently plantlets were transferred out of green house and planted in larger plastic pots containing loam soil with DAP fertilizer. The regenerants were studied morphologically with regard to general appearance, shoot number, length of shoot and number of roots formed in both conventional media in comparison to those of low cost media. Survival of plantlets was recorded after 3 weeks [Survival plantlet (%) = (Surviving plantlets/Total plantlets) x 100).

2.6 Data Analysis

Differences on the number of shoots, plant height and number of roots were subjected to analysis of variance (ANOVA). Means were separated by post hoc Tukeys at $p < 0.05$. The cost efficiency (CE) was calculated by dividing the price of low cost media substitutes by conventional media per litre then subtracting from 100%.

3. RESULTS AND DISCUSSION

3.1 Cost Analysis between the Low Cost Medium and the Conventional (MS) Medium

The same media composition was used during initiation and multiplication for the four treatments (conventional media, LCM1, LCM2 and LCM3). Therefore cost reduction achieved at 94% for LCM1 (Table 2) and 96% LCM3 and LCM4 (Table 3 and Table 4 respectively).

3.2 Shoots Regeneration in Various Media

Significant differences ($p < 0.05$) were observed in the number of shoots produced per plant among the treatments (Fig. 1). Dasheen had the highest number of shoots (5.92) on MS media followed by LCM3 (5.48), LCM2 (5.29) and LCM1 (4.7). The number of plant shoots produced in LCM1 was significantly lower number compared to LCM2, LCM3 and MS media.

No significant differences were observed on the growth of Dasheen, Eddoe and Wild/Purple plants on the MS media, LCM1, LCM2 and LCM3 (Fig. 2). Eddoe had the highest number of shoots (5.52) followed by Dasheen (5.5) and wild (5.05) on the media.

Table 1. Composition of each media tried

Media component		
Macronutrients	**Micronutrients**	**Media code**
Conventional	Conventional	Conventional
Omex	Omex	LCM1
Fertilizers	Stanes	LCM2
Fertilizers	Microfood	LCM3

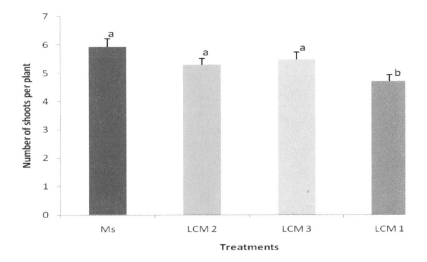

The same letter (s) expressed show no difference at P<0.05 level
Fig. 1. Shoot generation by use of different media

Table 2. Cost analysis of LCM1 compared to MS media

Conventional	Low cost	Cost of 1 litre of medium (Kenyan shillings)		Cost reduction
Macronutrients		**Conventional**	**Low cost**	
$MgSO4$		0.4810		
KNO_3		3.4200		
$CaCl_2$		0.7920	3.0000	
NH_4NO_3		4.9500		
KH_2PO_4		0.2890		
Sub-total		9.9320	3.0000	
Micronutrients				
$COCL_2.6H_2O$		0.0002		
$CuSO_4. 5H_2O$		0.0002		
Na_2EDTA		0.0154		
$FESO_4.7 H_2O$	Omex Foliar	0.0078		
H_3BO_3		0.0512		
KI		0.0035		
$MnSO4.4 H_2O$		0.0605		
$Na_2MoO_4 .2 H_2O$		0.0039		
$ZnSO_4.7 H_2O$		0.0008		
Sub-total		0.1436	3.0000	70.2200
Total		10.0756	3.0000	
Sucrose	Sugar	105.0000	3.0000	97.1000
Total		115.0756	6.0000	94.7860

Table 3. Cost analysis of LCM2 compared to MS media

Conventional	Low cost substitute	Cost of 1 litre of medium (Kenyan shillings)		Cost reduction %
Macronutrients		Conventional	Low cost	
MgSO4	Epsom salt	0.4810	0.0330	96.1670
KNO_3	Potassium Fert.	3.4200	0.1710	95.0000
$CaCl_2$	Calcinit	0.7920	0.7920	0
NH_4NO_3	Ammonia Fert.	4.9500	0.1897	95.3620
KH_2PO_4	MonoPotassium	0.2890	0.0204	92.9410
SUBTOTAL		9.9320	1.1951	· 95.8270
Micronutrients				
$COCL_2.6H_2O$		0.0003		
$CuSO_4. 5H_2O$		0.0002		
Na_2EDTA	Stanes	0.0154		
$FESO_4.7 H_2O$		0.0078		
H_3BO_3		0.0512	0.4049	
KI		0.0035		
$MnSO4.4 H_2O$		0.0605		
$Na_2MoO_4 .2 H_2O$		0.0039		
$ZnSO_4.7 H_2O$		0.0008		
Sub-total		0.1436	0.4049	
Total		10.0756	1.6000	
Sucrose	Table sugar	105.0000	3.0000	97.1000
TOTAL		115.0756	4.6000	96.0000

Table 4. Cost analysis of LCM3 compared to MS media

Conventional	Low cost substitute	Cost per litre of medium (Kenyan shillings)		Cost reduction (%)
Macronutrients		Conventional	Low cost	
MgSO4	Epsom salt	0.4810	0.0330	96.1670
KNO_3	Potassium Fer.	3.4200	0.1710	95.0000
$CaCl_2$	Calcinit	0.7920	0.7920	0
NH_4NO_3	Ammonia Fer.	4.9500	0.1890	95.3620
KH_2PO_4	MonoPotassium Phos	0.2890	0.0200	92.9410
Sub-total		9.932	1.1951	95.8270
Micronutrients				
$COCL_2.6H_2O$		0.0002		
$CuSO_4. 5H_2O$		0.0002		
Na_2EDTA		0.0154		
H_3BO_3	Micro-food	0.0512		
KI	horticulture	0.0034		
$MnSO4.4 H_2O$		0.0605		
$Na_2MoO_4 .2 H_2O$		0.0039		
$ZnSO_4.7 H_2O$		0.0008	0.4049	
Sub-total		0.1436		
Total		10.0756	1.6000	
Sucrose	Table sugar	105.0000	3.0000	97.1000
Total		115.0756	4.6000	96.0000

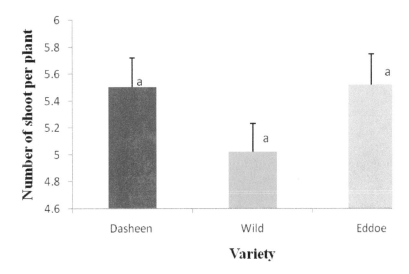

The same letter (s) expressed show no difference at P<0.05 level
Fig. 2. Shoot regeneration across the varieties

3.3 Height of the Regenerated Plants

The height of all the taro varieties was significantly different (p<0.05) in all the treatments. The three taro varieties grown on MS media were taller compared to LCM1, LCM2 and LCM3 (Fig. 4). Plants grown on LCM1 had the shortest heights ranging from 2.56 cm (Eddoe) to and 2.94 cm (Dasheen) compared to MS media that ranged from 6 cm (Dasheen) to 5.61 cm (wild) Fig. 5.

3.4 The Number of Roots and Transplant Survival Rate

Significant differences (p<0.05) were observed on the number of roots generated using 1mg/l (Indole Acetic Acid) IAA and Citishooter. IAA on Eddoe variety had the highest number of roots (6.56) compared to wild variety (5.56 cm) on LCM3 media. There was no significant variation on the *in vitro* plantlets survival rate after three weeks Fig. 7. The survival rates were 99.35% (MS media), 99.3% (LCM3) and 99.1% (LCM1)

As observed in Fig. 6. The tested media supported growth of shoots with the same morphology and plant structure as those of conventional media.

The LCM1 substitute for Ms Media resulted in cost reduction by 94.786% (Table 2) while LCM2 and LCM3 each further reduced cost by 96.0% (Table 3 and Table 4 respectively). This supports

previous studies by Gitonga et al. [4] who substituted macronutrients and micronutrients with the alternatives that reduced the cost by 94.2 and 97.8% respectively in banana micro-propagation. Ogero et al. [8] also studied how low cost nutrients can be substituted in cassava in which he was able to demonstrate that similar percentage of savings was achieved. This has further been shown by [9] low cost tissue culture of sweet potato. This reduction in the cost of media shows that the planting materials obtained can benefit resource poor farmers. The farmers will have access to planting materials at a cheaper price than those propagated using MS Media.

The substitute of plant culture media as a low-cost strategy to propagate planting material must guarantee high quality and well developed plants that compare well with conventional counterparts in both green house and field conditions. Multiple shoots were obtained after 40 days with transfer to new media after every 10 days (Fig. 3). The average number of shoots obtained for the control media were similar to those of Chien-Ying et al. [2] who recorded an average of 5.9 shoots per explants using 8 mg/l of BAP. In this study, the use of 10 mg/l of BAP produced an average of 6.44 shoots. The number of shoots obtained using low cost media was not significantly different except for LCM2 on wild variety and LCM1 for Eddoe variety. This is an indication that micropropation of taro using LCM3 can substitute MS media for production of all varieties of taro.

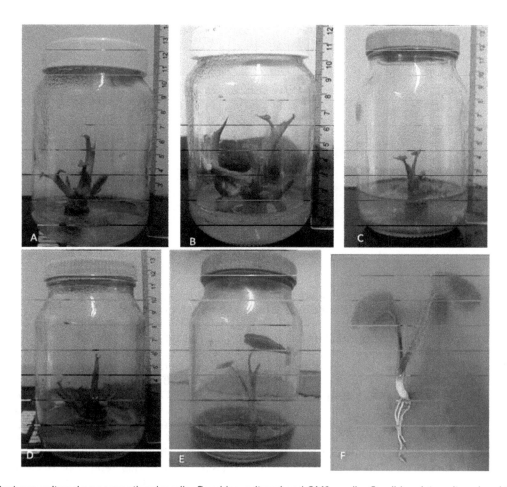

A: dasheen cultured on conventional media. B: eddoe cultured on LCM3 media. C: wild variety cultured on LCM2 media. D: dasheen variety cultured on LCM1 media. E: dasheen shoots rooted on media using citishooter as rooting hormone. F: rooted eddoe shoot on LCM3 media with citishooter as rooting hormone

Fig. 3. *In vitro* generated plants

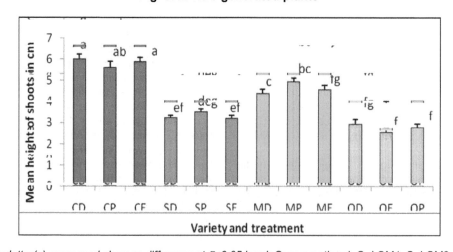

The same letter(s) expressed show no difference at P<0.05 level; C: conventional, O: LCM1, S: LCM2, M: LCM3; D: dasheen, E: eddoe, P: wild varieties of taro; means within a column followed by the same letter(s) are not significantly different at P<0.05

Fig. 4. Height of plants in treatment per variety

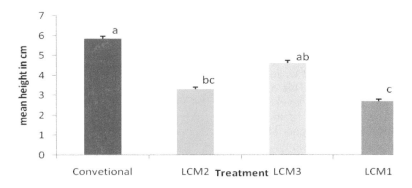

Same letter(s) expressed show no difference at P<0.05 level
Fig. 5. Height of plants per treatment

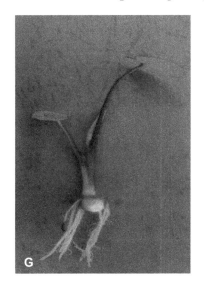
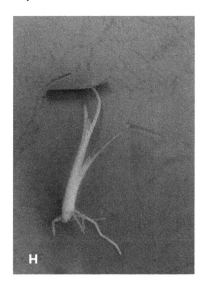

(G) Wild taro on conventional media.   *(H) Dasheen variety generated using LCM3 and citishooter*

Fig. 6. *In vitro* generated wild and dasheen varieties plants ready for hardening off

(I) In vitro taro plants on soil generated by conventional media a month after hardening off   *(J) In vitro taro plants generated on LCM3 a month after hardening off*

Fig. 7. Hardened *in vitro* generated taro

The height of plants varied with conventional media having average height of 5.83 cm, LCM1 2.7 cm, LCM2 4.63 cm and LCM3 3.31cm. This shows that the best substitute for production of tall plants is LCM2 while LCM1 produced the shortest plants. This can be attributed to the low phosphate levels in the Omex media. The optimum amount of phosphate in MS is 1.65 g/l compared to 0.48 g/l in the media used in this study. Higher concentrations of fertilizer is very toxic to the tissues and causes chlorosis to plants at the beginning and necrosis after several days of exposure [10]. There was no significance in root formation between use of 1 mg/l NAA and the low cost alternative 1 ml/l of citishooter. The root formation appeared after two weeks even though Dasheen and Eddoe varieties caused delay in rooting for 5 days. Roots have an essential role and function in plant life and development, supplying water and nutrients to the plant from the environment [11]. Acclimatisation and high percentage survival of plantlets shows the plants ability to withstand transplanting stress [12]. The high percentage of acclimatisation observed in this study may be attributed to plantlets with functional root system during *ex vitro* acclimatization [13]. Also the plantlets produced through micro-propagation technique were of high quality compared to those of vegetative means and vigorous with well developed leaves. The plants were also did adjust to the field conditions. The capability of plantlets to withstand *ex vitro* stress determines the success of any tissue culture protocol [7].

4. CONCLUSION

In this study the low cost alternative showed potential in the production of taro plantlets and should therefore be considered for adoption. This will ensure that there is availability of cheaper planting material to the farmers. This will contribute to food security as well as saving on the available resources especially in sub-Saharan Africa.

COMPETING INTERESTS

Authors have declared that no competing interests exist.

REFERENCES

1. Revill PA, et al. Incidence and distribution of viruses of taro (*Colocasia esculenta*) in Pacific Island countries. Australasian Plant Pathology. 2005; 34:327-331.

2. Chien-Ying K, Ji-Ping K, Rohan M. *In vitro* micropropagation of white dasheen (*Colocassia esculenta*). African Journal of Biotechnology. 2008;7(1):41-43.

3. Manner HI, Taylor M. Farm and Forestry Production and Marketing Profile for Taro (*Colocasia esculenta*). In: Elevitch CR, (ed.). Specialty Crops for Pacific Island Agroforestry. Permanent Agriculture Resources (PAR), Holualoa, Hawai'I; 2010.

4. Gitonga NM, et al. Low technology tissue culture materials for initiation and multiplication of banana plants. African Crop Science Journal. 2010;18(4):243-251.

5. Brink JA, Woodward BR, DaSilva EJ. Biotechnology: A tool for development in Africa. Electronic Journal of Biotechnology. 1998;1(3).

6. Savangikar VA. Plant tissue culture for entrepreneurs, business houses, farmers and nurserymen and natural/herbal products; 2002.

7. Ahloowalia BS, Savangikar VA. Low cost options for energy and labour. Proceedings of a Technical Meeting organized by the Joint FAO/IAEA Division of Nuclear Techniques in Food and Agriculture and held in Vienna, 26-30 August; 2002.

8. Ogero KO, et al. Cost-effective nutrient sources for tissue culture of cassava (Manihot esculenta Crantz). African Journal of Biotechnology. 2012; 11(66):12964-12973.

9. Ogero KO, et al. A low-cost medium for sweet potato micropropagation. African Crop Science Conference Proceedings. 2011;10:57-63.

10. Santana MA, et al. A simple and low-cost strategy for micropropagation of cassava (*Manihot esculenta* Crantz) African Journal of Biotechnology. 2009; 8(16):3789-3897.

11. Schiefelbein JW, Masucci JD, Wang H. Building a root: The control of patterning and morphogenesis during root

development. Plant Cell Rep. 1997; 9:1089-1098.

12. Ziv M. *In vitro* hardening and acclimatization of tissue cultured plants. In: Plant tissue culture and its agricultural applications. Withers LA, Alderson PG, (Eds.). Buttersworhts, London. 1998;187-203.

13. Demo P, et al. Table sugar as an alternative low cost medium component for *in vitro* micro-propagation of potato *(Solanum tuberosum L.)*. African Journal of Biotechnology. 2008;7(15):2578-2584.

9

Towards Efficient *In vitro* Regeneration of Cowpea (*Vigna unguiculata* L. Walp): A Review

Lawan Abdu Sani[1,2*], Inuwa Shehu Usman[1], Muhammad Ishiaku Faguji[1] and Sunusi Muhammad Bugaje[1]

[1]*Department of Plant Science, Ahmadu Bello University, Zaria, Nigeria.*
[2]*Department of Plant Biology, Bayero University Kano, Nigeria.*

Authors' contributions

This work was carried out in collaboration between all authors. Authors LAS and ISU did the literature searches, composed and wrote the first draft of the manuscript. Authors MIF and SMB edited the manuscript drafted by authors LAS and ISU. All authors read and approved the final manuscript.

Editor(s):
(1) David WM Leung, School of Biological Sciences, University of Canterbury, New Zealand.
(2) Giovanni DalCorso, Department of Biotechnology, University of Verona, Italy.
Reviewers:
(1) Ali Movahedi, Nanjing Forestry University, China.
(2) Anonymous, Université Cheikh Anta Diop, Dakar, Sénégal.

ABSTRACT

Cowpea is a crop of tremendous economic and ecological values particularly in sub-Saharan Africa, where over 80% of the crop is produced and consumed. Due to heavy attack by pests and diseases, actual yield does not exceed 20% of the crop's potential in most of the production regions. Shortages of genes to combat biotic stresses in the germplasm and sexual incompatibility with wild relatives are major impediments in cowpea improvement. Genetic modification of cowpea with relevant genes can address these problems. Establishment of reproducible regeneration system is a prerequisite for genetic transformation of cowpea using transgenic technology. Cowpea is among the most recalcitrant crops for manipulation under *In vitro* condition especially via *de novo* process. However, strategies to regenerate cowpea under *In vitro* conditions have evolved steadily in the last three decades. In this review, we give a summary of cowpea regeneration work carried out so far and discussed approaches employed as well as challenges of developing efficient regeneration systems in cowpea.

*Corresponding author: Email: sanimalk@yahoo.co.uk

Keywords: Cowpea; In vitro culture; organogenesis; somatic embryogenesis; efficient regeneration system.

1. INTRODUCTION

Among the genus *Vigna*, cowpea (*Vigna unguiculata*) and Bambara nut (*Vigna subterranea*) are the two species domesticated in Africa [1]. Cowpea is consumed in tropical and sub-tropical countries of Africa, Asia and America. It is an important source of dietary protein with both its dry grain and vegetative parts containing 23-32% protein and other important food components such as lipid, fiber and vitamins [2]. It complements protein deficient maize, cassava, millet and sorghum based staple foods consumed in sub- Saharan Africa. Cowpea occupies an estimated area of 14.5 million hectares throughout the world with annual production of 5.5 million tonnes [3]. Over 80% of cowpea production comes from West Africa with Nigeria, Niger and Burkina Faso accounting for 77% of the production [4]. It is cultivated as a security crop by small scale farmers due to its drought tolerance and ability to thrive well in low fertile soils. Despite its nutritional and economic importance, productivity of cowpea has not exceeded 20% of its potential with yield limited to an average of 0.37/t/ha in Africa [5].

Efforts to address production problems lead to the establishment of a number of cowpea breeding programs by many research centers around the world. Significant progress has been recorded in the development and release of varieties with tolerance to diseases, drought and parasitic weeds [6]. This achievement is in part, due to the identification of molecular markers for various traits leading to the development of tools for marker-assisted selection-based breeding [7]. However, conventional breeding has not provided adequate solution to the problems of insect pests and viruses. While cowpea generally lacks inherent defence against destructive insects, some levels of resistance to few viruses have been reported [8]. In most of these cases however, resistance has been classified as tolerance not inherent immunity in the plants [9].

Realization of the benefit of biotechnology application in Agriculture and the success recorded in crops like Cotton and Maize, made scientists to dedicate efforts in deploying modern scientific tools to develop cowpea varieties expressing genes for resistance against insects and viruses as well as other traits of interest. Transgenic technology therefore, represents an alternative means of introgression genes of interest into cowpea, thereby providing lasting solutions to production constraints. Success in genetic manipulation of cowpea and other crop plants depends on the availability of In vitro regeneration system that will provide totipotent cells capable of regenerating complete plants following gene delivery. In this review we reported the approaches employed in the development of reproducible protocols for In vitro regeneration of cowpea. We also reported the success made and highlights the prospects tissue culture holds in overcoming challenges of cowpea improvement.

2. COWPEA TISSUE CULTURE

Application of biotechnology approach to confront challenges of cowpea improvement represents an important alternative that may accelerate production of varieties with useful traits. This approach when applied alongside with conventional breeding will complement the efforts of breeders in overcoming challenges of cowpea production. Tissue culture represents a key biotechnology tool used in exploiting the plant totipotency concept proposed by Haberlandt [10] for the propagation of plants. In the last three decades efforts have been devoted to the development and optimization of regeneration systems in cowpea. The ultimate target is to develop reproducible tissue culture protocols that will serve as a basis to achieve stable genetic transformation and In vitro selection for traits of agronomic importance. This strategy allows for generation of elite cowpea lines expressing traits that are almost not possible to obtain by conventional breeding.

Although, there have been advances in the field of plant tissue culture, cowpea and many other cultivated legumes proved to be highly recalcitrant and are amenable to regeneration systems only after modification of tissue culture procedures [11]. The two major regeneration pathways; organogenesis and somatic embryogenesis have been employed in the In vitro regeneration of cowpea (Fig. 1).

2.1 Organogenesis

Organogenesis entails the initiation and development of shoots or roots either directly from organized tissue in the explant or from callus culture. One of the first reports of successful regeneration of cowpea via organogenesis was from Kartha et al. [12] This group reported regeneration of shoots from apical meristem of cowpea on hormone-free MS medium and media supplemented with a low concentration of N^6-benzyladenine (BA) (0.005-0.1 µM) in combination with 0.05 µM NAA . Following this report, scientists used different approaches including media combinations, different genotypes, different types of explants, as well as different types and combinations of hormones to improve regeneration efficiency via organogenesis (Table 1).

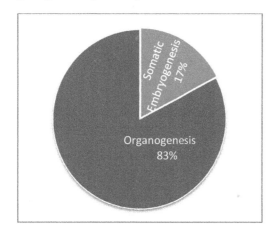

Fig. 1. Percentage distribution of morphogenic pathways in reported cases of successful regeneration of cowpea from 1981 to 2014 (Table 1)

2.2 Media and Other Additives

Following the report of Kartha et al. [13,14] it took a decade before a breakthrough was made in the development of cowpea culture media. The group developed a new basal medium (BM) for embryo rescue following analysis of mineral composition of embryos and MS medium. This was achieved by determining the mineral salts composition of embryos of *Vigna species* and autoclaved MS medium and based on the minerals composition of the embryos, nutrient composition of BM medium was formulated relative to that of MS medium. The new BM medium resulted in the increase in regeneration frequency from less than 20% to between 35-55% (Table 1). This finding and the report of [15] revealed the efficiency of combination of MS basal salts and B5 vitamins in the *In vitro* regeneration of cowpea leading to wide adoption of MSB (MS salts+B5 vitamins) formulation in

cowpea research. In the last three decades MSB, MS and B5 have been the most common formulations used in reported cases of successful regeneration of cowpea (Table 1).

In addition to normal constituents of the media, increase in organic nitrogen in culture medium was reported to have effect on cowpea regeneration. Addition of Casein hydrolysate or L-glutamine was reported to improve *In vitro* regeneration of cowpea [16]. The preferential utilization of organic nitrogen during morphogenesis in cowpea has been advanced as the possible reason for increased regeneration frequency with addition of organic nitrogen sources [13,14]. Another strategy through which improved regeneration of cowpea was achieved is the addition of ethylene synthesis and action inhibitors. Incorporation of compounds like $COCl_2$, $AgNO_3$ and 2, 5-norbornadiene in the culture media was reported to improve shoots regeneration in cowpea [17]. This could be explained by the fact that ethylene is released by actively dividing cultured cells and has been reported to promote callus growth and suppress shoots development under *In vitro* condition. Therefore, addition of ethylene synthesis and ethylene action inhibitors to cowpea culture media will be accompanied by expected increase in regeneration frequency.

2.3 Genotype and Explants

Genotype proved to be one of the significant factors that affect regeneration capacity in cowpea. The importance of genotype in cowpea regeneration was first demonstrated by work of [18]. The authors regenerated less than 50% of the 36 US cowpea genotypes they evaluated by initiating cotyledons on induction medium containing 1/3 MS fortified with 66.6-155.4µ M/L BA and subsequent shoots regeneration on MS supplemented with 4.44µ M/L BA. In addition to the differences in regeneration capacity among the genotypes, a great variation in regeneration frequencies (1 to11%) and number of shoots per explant (2 to 14) were also observed among the regenerated genotypes. Recently [19-22] have also reported significant influence of genotype on *In vitro* regeneration of cowpea. In all the reports, cowpea genotypes not only differ in their regeneration potentials, but also in the degree to which they responded to prevailing culture conditions. The variations might be due to both genetic and physiological differences among the genotypes, which underscore the need for genotype-dependent regeneration protocol in cowpea.

Although genotype alone constitutes a major source of variation in cowpea regeneration, the specific tissue chosen for regeneration is also an important consideration. Morphogenic response in cowpea was reported to be governed by selection of suitable explant [17,23,24]. Several explant types which included mature and immature embryos, primary leaves, cotyledon, cotyledonary node, epicotyl, hypocotyl, thin cell layer, stem segment, shoot tip, plumular apices, shoot meristem and root tip have been employed in the regeneration of cowpea (Table 1). In almost all cases of successful cowpea regeneration, explants with active merismatic tissue were reported to exhibit better regeneration ability. This is because merismatic tissues are known to be the sources of regenerative potential of plants due to the presence of undifferentiated cells with genetic capability to generate complete plants. The presence of merismatic cells therefore, play a key role in cowpea's ability to regenerate under In vitro condition and selection of explant with sufficient merismatic tissue is one of the key steps towards successful regeneration of cowpea.

2.4 Types and Concentrations of Hormones

In cowpea organogenesis, the degree of variability in regeneration frequency and time required to achieve regeneration indicate a wide range of activities of growth regulators. Concentrations and combinations of growth regulators constitute some of the major differences from one report to another in the In vitro regeneration of cowpea. Among the categories of growth regulators, cytokinin type and concentration appear to be the most important in cowpea organogenesis. The most widely employed cytokinin types are N^6-benzylaminopurine (BAP), kinetin (KT) Thidiazuron (TDZ) and Zeatin (ZE) (Table 1).

N^6-benzylaminopurine (BAP) is the predominant cytokinin present in most of the reports of cowpea regeneration and its efficacy has been demonstrated [21,25-32]. In most of these reports, maximum shoot induction were achieved using different levels of BAP. This might be due to genotype-dependent physiological response to BAP during the induction stage. Generally BAP levels ranging from 0-10 µM (Table 1) were reported to be suitable for shoot regeneration of cowpea. The optimum concentration however, depends on the genotype and explant used.

In reports where BAP was not used, alternative cytokinins such as TDZ, KT and ZE were incorporated in the generation media. TDZ has been reported to induce multiple shoots production from cowpea thin cell layer [33], cotyledonary nodes [27,34] and shoot apices [20], as well as leaves, stem segments and cotyledons [35]. In these reports TDZ has been shown to be potent in inducing shoots morphogenesis in concentrations within the range of 0.68-1.6 µM. At higher concentration however, TDZ inhibits shoot elongation as reported by [28]. TDZ is an important modulator of adventitious shoots formation under In vitro condition and has proved to be useful in species considered to be recalcitrant. It is reported to act rather indirectly by regulating the activities of endogenous hormones [36]. The physiological and morphogenic effects of KT, ZE and GA_3 have also been studied in cowpea under In vitro condition (Table 1). ZE has been reported to be very effective in improving morphogenesis and regeneration in cowpea [13,14]. Although KT is an active cytokinin and has been reported to induce morphogenesis in many plant species, its physiological role in cowpea was reported to be similar to that of GA_3. Both KT and GA_3 were reported to promote shoot elongation in cowpea [26,27,35].

It is known that the ratio between auxin and cytokinin is crucial in achieving regeneration under In vitro condition. Auxin-cytokinin ratios of induction media [37] and in the explant, determine either shoot or root regeneration. The most common auxins used in cowpea regeneration via organogenesis are Indole-3-butyric acid (IBA), Indoleacetic acid (IAA), naphthalene acetic acid (NAA) and 2, 4-dichlorophenoxyacetic acid (2, 4-D). Apart from the pioneering work on cowpea embryo rescue and regeneration via organogenesis from cotyledon and hypocotyl explants and the recent report of preconditioning of embryonic axes in NAA [38], no attempt is reported on evaluating the efficacy of auxins as sole hormones in regeneration of cowpea via organogenesis. While pellegrinesch et al. [13,14] reported IAA and NAA to effectively suppress shoot development, Aasim reported increase in shoot morphogenesis following preconditioning in NAA and subsequent regeneration on MS fortified with a low concentration (1.11-4.44 µM/L) BAP. The author, however, reported a deleterious effect of incubation in NAA for a long period which resulted in browning and subsequent death of the explant. Auxins were reported to increase regeneration in cowpea only in the presence of cytokinins. Auxins are, however, known to promote rhizogenesis at optimum concentrations and have been reported to increase root formation under In vitro condition in cowpea [16,26,34,39].

Plant regeneration in cowpea was reported from both untreated explants and explants treated by preconditioning in a high concentration of cytokinin. Shock treatment with a high concentration of cytokinin appears to trigger plant development in cowpea, leading to differentiation and subsequent plantlet development. The strategy has been employed to increase shoot morphogenesis in cowpea by preconditioning different explants in BAP [18,23,29,30] and TDZ [34].

Table 1. Historical development of tissue culture in cowpea (*Vigna unguiculata* L.) from 1981 to 2014, indicating the frequencies of morphogenic pathways, explants types, media compositions, plant growth regulators and progress made in regeneration efficiency

Explants type	Mode of regeneration	Medium	PGRs	Regeneration frequency	Ref.
Shoot apical meristems	Organogenesis	MS	BAP, NAA	1-11%	[12]
Immature cotyledons	Somatic embryogenesis	MS MSB	BAP,2,4-D, NAA	10-15%	[15]
Primary leaves	Somatic embryogenesis	MS	BAP,2, 4-D	-	[41]
Primary leaves	Organogenesis	CS23, MS, SH23	BAP, 2,4,5-T, NAA, IAA, 2,4-D	20 %	[25]
Epicotyls, cotyledons and hypocotyls	Organogenesis	MS	BAP, IAA	15%	[23]
Shoot tip	Organogenesis	MS	BAP, NAA, 2,4-D	Not reported	[16]
Hypocotyls and cotyledons	Organogenesis	BM, MSB	BAP, ZE,KT IAA, NAA	36-55%	[14]
Cotyledons	Organogenesis	MS	BAP	1-11%	[18]
Primary leaves	Somatic embryogenesis	MS	2,4-D, ABA	21-28%	[42]
Cotyledonary thin cell layer explants	Organogenesis	MSB	TDZ, IBA	80%	[33]
Primary leaves	Somatic embryogenesis	MSB, B5	2,4-D, ABA	32%	[43]
Shoot apices epicotyls, stem	Organogenesis	MSB	BAP, GA3, IBA	77%	[17]
Hypocotyls, epicotyls, stem segments, leaves and terminal	Organogenesis	MS	BAP,NAA, 2,4-D	-	[24]
Cotyledonary nodes	Organogenesis	MSB, MS	BA, IBA	-	[39]
Cotyledonary nodes	Organogenesis	MS B5	BAP, KT, NAA	100%	[26]
Shoot meristem	Organogenesis	MS	BAP	76-87%	[19]
Cotyledonary nodes	Organogenesis	MSB	BAP	Not reported	[29]
Shoot meristem	Organogenesis	MS	BAP, NAA, IBA	55-67%	[20]
Shoot tip explants	Organogenesis	MS	TDZ, IBA	100%	[28]
Plumular apices	Organogenesis	MS	BAP, NAA, IBA	8-100%	[38]
Cotyledonary nodes	Organogenesis	MSB	BAP	-	[32]
Cotyledonary nodes	Organogenesis	MSB	BAP	95%	[30]
Cotyledonary nodes	organogenesis	MSB	BAP, KT	-	[31]

Key: PGRs - plant growth regulators, Ref - references

These studies reported the use of raised concentrations of cytokinins in the induction media to stimulate high frequency shoots formation. The induction phase is then followed by appropriate culture manipulations such as subculture on cytokinin-free media or media containing low concentration of cytokinin to achieve vigorous shoots development. The reports indicated that "hormone pulse" is important in cowpea especially during the induction stage. However, exposure of the explants to the plant growth regulator beyond the induction period was reported to have negative consequences on both shoot induction and plantlet development [13,14,38].

3. SOMATIC EMBRYOGENESIS

Somatic embryogenesis is the emergence of embryo-like-structures which can develop into complete plants, either directly from genetically predetermined cells in the explant or from intervening callus through a process similar to zygotic embryogenesis. Formation of embryogenic cell from non-embryogenic somatic cell results from unequal division of the progenitor cell to produce small embryogenic cell with dense cytoplasm and large vacuolated cell. The embryogenic cell then divides in an irregular pattern to form proembryonic mass or divides in a highly organized pattern to form somatic embryos [40]. In the presence of induction medium, somatic embryos bypass the normal developmental pattern to undergo direct secondary somatic embryogenesis or precocious germination to plantlets [40]. Two important factors are crucial in plant regeneration via somatic embryogenesis: (i) selection of explants (explant in the appropriate developmental stage) that will provide competent progenitor cells and (ii) development of appropriate medium combination that will provide enabling environment for somatic embryogenesis. Although reports dealing with somatic embryogenesis in cowpea are scarce, these two factors have to some extent been investigated in the few reports available so far. Regeneration via somatic embryogenesis was reported from immature embryo [15] and primary leaf explants [41-43]. These explants were reported to provide progenitor cells from which embryogenic callus and subsequently somatic embryos were regenerated.

Somatic embryogenesis and plant regeneration in cowpea are influenced by components of culture medium which include auxin, cytokinins, sources of organic nitrogen, stress inducing factors and stress response factors. Addition of compounds such as 2,4-D, BAP, ZE, TDZ, KT, casein hydrolysate, L-glutamine, mannitol, ABA and L-proline to the culture media were found to enhance somatic embryogenesis in cowpea [15,41-43].

The strategy employed entails induction of embryogenic callus by exposing explants to a higher concentration of 2, 4-D (6.78-9.05 µM) followed by a gradual decrease in 2, 4-D concentration during development (1.13-2.26 µM) and maturation (0.05-0.45 µM) of somatic embryos [42,43]. In somatic embryogenesis, the presence of auxins is not only important during callus induction but also plays a significant role albeit at a low concentration during embryo development. Decreases in 2, 4-D and IAA levels were shown to be associated with early and later stages of embryo development in carrot [44]. To achieve efficient maturation of cotyledonary stage somatic embryos, stress was induced by addition of 2-3% mannitol and to enable somatic embryos respond to the stress induced, stress response factors such as ABA (5 µM)and L-proline (173.7 µM) were added to the regeneration media [43]. The role of organic nitrogen sources in the development and maturation of somatic embryos in cowpea was reported [43]. The beneficial effect of casein hydrolysate and L-glutamine could be explained by preferential utilization of organic nitrogen sources during seed development in cowpea [13,14]. During seed maturation in cowpea, nitrogen is translocated directly from leaflet protein reserves to developing seeds and not from the symbiotic nitrogen-fixing activity in the roots, as is the case in other legumes [45].

Conversion of cotyledonary somatic embryos to plantlets was enhanced by addition of ZE [42] or TDZ [43]. Cytokinin plays two key roles in germination of somatic embryos, (i) breaking dormancy induced by incubation in ABA and (ii) ensuring proper differentiation of shoot apices in somatic embryos. In the absence of cytokinin, plantlet recovery was reported to be low due to poor differentiation of shoot apices caused by prolonged exposure to auxin.

4. CONCLUSION AND PROSPECTS

Cowpea is a crop of tremendous economic and ecological values in sub-Saharan Africa where over 80% of the crop is produced and consumed. Cowpea is among the most recalcitrant crops for manipulation under In vitro condition especially via *de novo* processes [11]. Establishment of a reproducible regeneration system is a prerequisite for genetic transformation of cowpea using transgenic technology. Strategies to regenerate cowpea under *In vitro* condition have evolved steadily in the last three decades. It is now clear that regeneration of cowpea via organogenesis is well established and has been utilized as a base upon which transformation protocols are developed. However, this system demonstrated low regeneration frequency of transgenic plants which are also chimeric in nature due to the multicellular origin with poor transmission of transgenes to progenies. Despite these challenges significant progress has been made, transgenic lines expressing genes of interest are now grown under containment and field trials in different parts of the world [7]. Strategies to improve the frequency of plantlets regeneration are needed in order to increase the efficiency of cowpea regeneration via organogenesis. In an experiment conducted in our laboratory to evaluate the effect of different concentrations of BAP on shoot induction, we observed that, in the presence of BAP, a large number of buds was induced from the apical region of embryonic axes. However, only few (3 to 7) buds actually developed into fully elongated plants (data unpublished). Poor plantlet development following bud induction might be among the possibilities that explain low plant regeneration in cowpea. A strategy to improve development of buds into complete elongated plants could increase the efficiency of cowpea regeneration based on *de novo* organogenesis, thereby increasing the chance of generating more transformed plants.

Assessment of the trend of studies on somatic embryogenesis in cowpea indicated some success, but mostly confronted with challenges yet to be overcome. Somatic embryogenesis remains the most vital strategy to generate convenient targets for gene delivery. Development of an efficient and reproducible somatic embryogenesis system holds the key to stable transformation of cowpea because of the single cell origin of somatic embryos. In the last two decades, only four reports of successful regeneration of cowpea via somatic embryogenesis were published. An important challenge common to all the reports is the low conversions of somatic embryos to plantlets (20-30%). More so, for almost a decade now, there has not been any new report on somatic embryogenesis in cowpea or indeed validation of the earlier reported protocols. More research is

therefore needed to improve on the existing protocols, especially in strategies of converting somatic embryos to plantlets. Identification of genes involved in somatic embryogenesis and adequate knowledge on hormonal interactions in signal transduction pathways during somatic embryogenesis will provide valuable information that will help scientist to break the recalcitrant jinx in cowpea. Keeping in view the current challenges in transformation of cowpea, an efficient somatic embryogenesis system can fill the missing gap and the future may see cowpea varieties expressing vital agronomic traits.

COMPETING INTERESTS

Authors have declared that no competing interests exist.

REFERENCES

1. Ba FS, Pasquet RS, Gepts P. Genetic diversity in cowpea [*Vigna unguiculata* (L.) Walp] as revealed by RAPD markers. Genetic Resources and Crop Evolution. 2004;51:539-550.

2. Nielson SS, Brandt WE, Singh BB. Genetic variability for nutritional composition and cooking time of improved cowpea lines. Crop Sci. 1993;33:469-472.

3. CGIAR. Cowpea. Available:www.cgiar/our-reseerch/crop.factsheet/cowpea (Accessed 23rd November; 2013)

4. FAOSTAT; 2012. Available:http://faostat3.fao.org/faostatgateway/go/to/download/Q/QC/E

5. Waddington SR, Li X, Dixon J, Hyman G, de Vicente MC. Getting the focus right: Production constraints for six major food crops in Asian and African farming systems. Food security. 2010;2:27-48. DOI 10.1007/s12571-010-0053-8.

6. Singh BB, Awika J. Breeding high yielding cowpea varieties with enhanced nutritional and health traits. Fifth world cowpea research conference, Sally-Senegal; 2010.

7. Citadin CT, Abdulrazak BI, Aragao FJL. Genetic engineering in cowpea (*Vigna unguiculata*) history, status and prospects. GM Crops. 2011;2(3):1-6. DOI: 10.4161/gmcr.2.3.18069.

8. Hampton RO, Thottappilly G. Cowpea. In: Loebenstein G, Thottappilly G, editors. Virus and virus-like diseases of major crops in developing countries. Kluwer Acad. Pub. 2003;355-376.

9. Loebenstein G, Katis N. Advances in virus research; control of plant virus diseases: seed-propagated crops. Academic press limited. 2014;461-463.

10. Haberlandt G. Kulturversuche mit isollierten pflanzenzellen. Weisen Wien Naturwissenschaften. 1902;111:69-92. (German).

11. Aragao FJL, Campos FAP. Common bean and cowpea. In: Pua EC, Davey MR, cditora. Biotechnology in agriculture and forestry. Transgenic crops IV. Berlin: Springer. 2007;263-76.

12. Kartha KK, Pahl K, Leung NL. Plant regeneration from meristems of grain legumes: Soybean, cowpea, peanut, chickpea and bean. Can. J. Bot. 1981;59: 1671-1679.

13. Pellegrineschi A, Fatokun CA, Thottappilly G, Adepoju AA. Cowpea embryo rescue: 1. Influence of culture media composition on plant recovery from isolated immature embryos. Plant Cell Rep. 1997;17:133–138. DOI: 10.1007/s002990050366.

14. Pellegrineschi A. In vitro plant regeneration via organogenesis of cowpea (Vigna unguiculata L.). Plant Cell Rep. 1997;17:89-95. DOI: 10.1007/s002990050358.

15. Li XB, Xu ZH, Wei ZM, Bai YY. Somatic embryogenesis and plant regeneration from protoplasts of cowpea (Vigna sinensis). Acta Botnica Sinica. 1993;35: 632-636.

16. Brar MS, Al-Khayri JM, Shambein CE, Mcnew RW, Morelock TE, Anderson EJ. In vitro shoot tip multiplication of cowpea Vigna unguiculata (L.) Walp. Journal of the Arkansas Academy of Science. 1997;51:41-47.

17. Mao JQ, Zaidi MA, Arnason JT, Altosaar I. In vitro regeneration of Vigna unguiculata (L.) Walp. cv. Blackeye cowpea via shoot organogenesis. plant cell tiss. Org. Cult. 2006;87:121-125. DOI: 10.1007/s11240-006-9145-8.

18. Brar MS, Al-Khayri JM, Morelock TE, Anderson EJ. Genotypic response of cowpea (Vigna unguiculata L.) to In vitro regeneration from cotyledon explants. In Vitro Cell. Dev. Biol. Plant. 1999;35:8-12. DOI:10.1007/s11627-999-0002-4.

19. Manoharan M, Khan S, James OG. Improved plant regeneration in cowpea through shoot meristem. J. App. Hortic. 2008;10(I):40-43.

20. Aasim M, Khawar KM, Ozcan S. Comparison of shoot regeneration on different concentrations of thidiazuron from shoot tip explants of Cowpea on gelrite and agar containing medium. Not Bot Hort Agrobo. 2008;37(I):89-93.

21. Bakshi S, Sahoo B, Roy NK, Mishra S, Panda SK, Sahoo L. Successful recovery of transgenic cowpea (Vigna unguiculata) using the 6-phosphomannose isomerase gene as the selectable marker. Plant Cell Rep. 2012; 31:1093-1103. DOI 10.1007/s00299-012-1230-3.

22. Sawardekar SV, Jagdale VK, Bhave SG, Gokhale NB, Sawardekar SV, Lipne KA. Genotypic difference for callus induction and plantlet regeneration in cowpea [Vigna ungniculata (L.) Walp]. International Journal of Applied Biosciences. 2013;1(1): 1-8.

23. Amitha K, Reddy TP. Regeneration of plantlets from different explants and callus cultures of cowpea (Vigna unguiculata L.). Phytomorphology. 1996;46(3):207-211.

24. Bao YI, Bai Y, Wang YM, Huang YY, Xu XJ, Xu QW. The regeneration of cowpea (Vigna unguiculata L.). Agric. Sci. Guang Dong. 2006;4:31-33.

25. Muthukumar B, Mariamma M, Gnanam A. Regeneration of plants from primary leaves of cowpea. Plant Cell Tiss. Org. Cult. 1995;42:153-155. DOI: 10.1007/BF00034232.

26. Diallo MS, Ndiaye A, Sagna M, Gassama-Dia YK. Plants regeneration from African cowpea (Vigna unguiculata L.) variety. Afr. J. Biotechnology. 2008;16:2828-2833.

27. Solleti SK, Bakshi S, Sahoo L. Additional virulence genes in conjunction with efficient selection scheme and compatible culture regime enhance recovery of stable transgenic plants in cowpea via Agrobacterium tumefaciens-mediated transformation. J. Biotechnol. 2008;135:97-104. DOI: 10.1016/jjbiotec.2008.02.008.

28. Aasim M, Khawar KM, Ozcan S. In vitro micropropagation from plumular apices of Turkish cowpea (Vigna unguiculata L.) cultivar Akkiz. Sci. Hortic. 2009;122:468-471. DOI: 10.1016/j.scienta.2009.05.023.

29. Raveendar S, Premkumar A, Sasikumar S, Ignacimuthu S, Agastian P. Development of a rapid, highly efficient system of organogenesis in cowpea (Vigna unguiculata (L.). Walp). S. Afr. J. Bot. 2009;75:17-21.

30. Manman T, Qian L, Yanxia Z, Huanxiu L.

Effect of 6-BA on the plant regeneration via organogenesis from cotyledonary Node of cowpea (*Vigna unguiculata* L. Walp). Journal of Agricultural Science. 2013;5(5):2013. DOI: 10.5539/jas.v5n5pl.

31. Negawa AT, Hassan F, Jacobsen HJ. Regeneration and *Agrobacterium* mediated transformation of Cowpea. Annual conference on tropical and subtropical agriculture and natural resources management (TROPENTAG). Czech University of Life Sciences Prague, Czech Republic. 2014;17-19.

32. Tang Y, Chen L, Li XM, Li J, Luo Q, Lai J, Li HX. Effect of culture conditions on the plant regeneration via organogenesis from cotyledonary node of cowpea (*Vigna unguiculata* L. Walp). African Journal of Biotechnology. 2012;11(14):3270-3275. Available:http://dx.doi.0rg/l 0.5897/AJB11.3214

33. Le BV, Carvalho MH, Zuily-Fodil Y, Thi AT, Van KTH. Direct whole plant regeneration of cowpea [(*Vigna unguiculata* (L.) Walp.] from cotyledonary node thin cell layer explants. J. Plant Physiol. 2002;159:1255-1258. DOI: 10.1078/0176-1617-00789.

34. Bakshi S, Sahoo L. How Relevant is recalcitrance for the recovery of transgenic cowpea: Implications of selection strategies. Journ. Plant Growth Regul. 2013;32:148–158. DOI: 10.1007/s00344-012-9284-6.

35. Li XM. Establishment of *In vitro* high efficient regeneration system of cowpea (*Vigna unguiculata* L. Walp.) and screening test for its resistance to kanamycin. Sichuan Agricultural University, China; 2011.

36. Murthy BNS, Singh RP, Saxena PK. Induction of high-frequency somatic embryogenesis in geranium (*Pelargonium x hortorum* bailey cv. ringo rose) cotyledonary cultures. Plant Cell Rep. 1996;15:423-426. DOI: 10.1007/BF00232068.

37. Che P, Gingerich D, Lall S, Howell S. Global and hormone induced gene expression changes during shoot development in arabidopsis. Plant Cell. 2002;14:2271-2279. DOI: 10.1105/tpc.006668.

38. Aasim M. *In Vitro* Shoot Regeneration of NAA-pulse Treated plumular leaf explants of cowpea. Not Sci. Biol. 2010;2(2):60-63.

39. Chaudhury D, Madanpotra S, Jaiwal R, Saini R, Kumar AP, Jaiwal PK. *Agrobacterium tumefaciens*-mediated high frequency genetic transformation of an Indian cowpea (*Vigna unguiculata* L. Walp.) cultivar and transmission of transgenes into progeny. Plant Sci. 2007;172:692-700. DOI: 10.1016/j.plantsci.2006.11.009.

40. Yang Y. Theories Behind Plant Tissue Culture-Somatic Embryogenesis plant and Soil Sciences e-library; 2012. Available:www.passel.unl.edu/pages/infor mationmodule.php (Accessed 12 December, 2014)

41. Kulothungan S, Ganapathi A, Shajahan A, Kathiravan K. Somatic embryogenesis in cell suspension culture of cowpea [*Vigna unguiculata* (L.) Walp], Israel J. Plant Sci. 1995;43:385-390.

42. Anand RP, Ganapathi A, Anbazhagan VR, Vengadesan G, Selvaraj N. High frequency plant regeneration via somatic embryogenesis in cell suspension cultures of cowpea (*Vigna unguiculata* (L.) Walp). *In vitro* Cell. Dev. Biol. Plant. 2000;36:475-480. DOI: 1054-5476/00.

43. Ramakrishnan K, Sivakumar PR, Manickam A. *In vitro* somatic embryogenesis from cell suspension cultures of cowpea (*Vigna unguiculata* (L.) Walp). Plant Cell Rep. 2005;24:449-461. DOI: 10.1007/s00299-005-0965-5.

44. Michalezuk L, Cooke TJ, Cohen JD. Auxin levels at different stages of carrot somatic embryogenesis. Phytochemistry. 1993;3:1097-1103.

45. Pate JS. Accumulation of seed reserves of nitrogen. In: DR Murray, editor. Seed physiology. Academic Press North Ryde. 1984;41–81.

10

Comparing the Effect of Two Promoters on Cassava Somatic Embryo at Transient GUS Assay Level

Olufemi O. Oyelakin[1,2*], Jelili T. Opabode[3] and Emmanuel O. Idehen[4]

[1]*Biotechnology Centre, Federal University of Agriculture, Abeokuta, Nigeria.*
[2]*Central Biotechnology Laboratory, International Institute of Tropical Agriculture (IITA) Ibadan, Nigeria.*
[3]*Department of Crop Production and Protection, Obafemi Awolowo University, Ile-Ife, Nigeria.*
[4]*Department of Plant Breeding and Seed Technology, Federal University of Agriculture, Abeokuta, Nigeria.*

Author's contribution

This work is a portion of the M.Sc. Dissertation of the author OOO. The other authors read and approved the final manuscript.

<u>Editor(s):</u>
(1) Giovanni DalCorso, Department of Biotechnology, University of Verona, Italy.
<u>Reviewers:</u>
(1) Ali Movahedi, Nanjing Forestry University, China.
(2) Anonymous, Nigeria.

ABSTRACT

35S promoter from the Cauliflower Mosaic Virus (pCaMV) is a constitutive promoter commonly used in plant genetic transformation while *Cassava Mosaic Virus* (pCsVMV) is another promoter which is underutilized. The combination of the two promoters was used to form (pOYE153). The method adopted includes the insertion of a β–glucuronidase reporter gene (UidA) into a promoter cassette comprising the CsVMV promoter. The second construct (pCAMBIA2310) had (pCaMV) used for the selectable marker and gene of interest. This construct was mobilized into *Agrobacterium tumefaciens* strain LBA4404 and then tested for expression of the UidA gene in transient assays in cassava somatic embryos. After co-cultivation of these *Agrobacterium* with the plant tissues, histochemical β–glucuronidase (GUS) assays were performed to determine the level of UidA gene expression in transient assays. The results showed that the pCsVMV was able to drive high gene expression of β–glucuronidase reporter gene (UidA) in the transient assays in cassava somatic embryo. Expression of the gene also increases with the increase in the day of co-cultivation and likewise expression of the gene was higher for the sample in the light than the dark.

Corresponding author: Email: olufemioyelakin@yahoo.com

Keywords: Cassava; promoter; pCsVMV; p(OYE153); p(CAMBIA2310) somatic embryo; beta-glucuronidase.

1. INTRODUCTION

Promoters are regions of the DNA that are located upstream of a coding region and which have specific sequences, recognized by proteins involved in the initiation of transcription of DNA to mRNA [1]. Promoters are important in the control of overall expression profile of a gene; it drives transcription at appropriate times and places. The basal promoter is located between 20-40 base pairs upstream of the start of the transcription and the promoter usually extends to about 200 base pairs or more upstream. The different types of promoters include constitutive promoters, tissue and cell-specific promoters, inducible and synthetic promoters. Promoters are used in the development of plant products and in research setting, so the type of promoter should be selected based on the application and desired expression pattern of the transgene in the plant [2]. Promoters comprise a set of transcription control modules clustered around the initiation site of RNA polymerase II [3].

The type of constructs and area of gene expression will determine the type of promoter to be used; some promoters express well in green tissue, root or reproductive tissues of the plant i.e. anther and ovary promoters that can confer tissue specific or temporal expression. Plant and viral promoters that drive high, constitutive expression have become valuable tools in plant genetic engineering. Among the noticeable promoters frequently used in plant genetic transformation is the *Cauliflower Mosaic Virus* 35S promoter CaMV35S [4].

Plant genetic transformation is a powerful means of studying gene expression in plants. It also allows for manipulation of biochemical processes that cannot be easily manipulated through conventional breeding. Genetic transformation is the heritable change in a cell or organism brought about by the uptake and establishment of introduced DNA. *Agrobacterium tumefaciens* is a plant pathogenic bacterium, which has become the most frequently used agent for the introduction of foreign genes into plant cells and the subsequent regeneration of transgenic plants. *Agrobacterium tumefaciens* naturally infects the wound sites in dicotyledonous plant causing the formation of the crown gall tumors.

The GUS reporter system (*GUS*: beta-glucuronidase; *uid*A) is a reporter gene system, particularly useful in plant molecular biology. Gene silencing refers to the inactivation of a transgene in a plant or animal cell. Gene silencing can occur at the transcriptional or post-transcriptional level. Gene silencing has also been called "homology-dependent" [5] or "repeat Induced" gene silencing [6]. It can involve the interaction between two unlinked loci in which one transgene locus is capable of trans-inactivating the second [7].

Gene silencing is found in plants and animals and is responsible for many important functions like defense against foreign organisms such as viruses [8]. Gene silencing can be caused by the presence of homologous sequences in a genome, for example, duplication of a gene in the genome can result in gene silencing. The duplication of promoter sequences has been shown to result in silencing of the genes controlled by these promoters [5]. The 35S promoter of the *Cauliflower Mosaic Virus* (CaMV) is considered a constitutive promoter and is widely used as the promoter for the marker gene and the gene of interest.

2. MATERIALS AND METHODS

2.1 Plasmids Description

The first plasmid was (pOYE153) comprises of pCaMV 35S used to drive the selectable marker while pCsVMV was used drive the gene of interest, the two promoters operated in opposite direction. The second plasmid was (pCAMBIA2301) where the selectable marker and the gene of interest were driven by pCaMV 35S and it was in the same direction.

2.2 Preparation of Cassava Somatic Embryo

Nodes were removed from in-vitro cassava plants and cultured on meristem enlargement medium for a week. The meristem was removed and it was placed on primary induction medium which contains picloram (10 mg/L) for 10 days in the dark so as to induce somatic embryo, after which the induced embryo was moved to secondary induction medium also in the dark for another 2.5 weeks [9,10]. The embryo matured after 17 days and it thereafter moved to maturation medium in the light for 2 weeks (Fig. 1).

2.3 Preparation of the Two Infection Media

The two bacteria constructs were inoculated into 3 mls YEB$_{Rif50 \ Sm300Km25}$ overnight and from the 3 mls culture, 1 ml solution was taken to inoculate 20mls YEB$_{Rif50 \ Sm300Km25}$. The next day, the bacteria grew and it was ready for infection. The bacteria constructs were allowed to grow to Optical Density OD$_{600}$ of about 0.8. They were later spun at 5,000 rpm for 15 mins and redissolved in YEB$_{Rif50 \ Sm300Km25}$ with 5 mM Acetosynringone and grown for another 2 hrs so that the OD increased to about 1.0 - 1.2. The constructs were spun and redissolved in plant medium with OD diluted to 0.5 for transformation [11].

2.4 Transformation of Cassava Somatic Embryo

The matured somatic embryos were harvested and it was chopped into smaller pieces for transformation. The somatic embryos were added into the diluted bacteria in the plant medium. The mixture was then placed on orbit shaker for 45 mins – 60mins so as to allow infection to take place very well. The somatic embryos were taken out and blotted on sterile paper for 5-10mins, and it was transferred to co-cultivation medium. The infected plant tissue on co-cultivation was placed at 22°C for 4 days. Transfer of T-DNA is optimal at 22°C – 24°C, some were placed in the dark and the rest in the light. After 2, 4 and 6 days histochemical GUS assay was done on the infected materials in the dark and in the light [11].

2.5 Histochemical Gus Assay

The histochemical GUS assay was done using [12] method. The assay was done somatic embryo. Few tissues of the infected somatic cotyledons of cassava were placed in 200 ul GUS solution, vacuum infiltrated for 1 minute, and incubated at 37°C for 4-8 hrs. Then the GUS solution was removed and chlorophyll in the tissues was removed with repeated 70% Ethanol washes.

3. RESULTS AND DISCUSSION

3.1 Expression of GUS Gene Using pCAMBIA2301

Genetic transformation of cassava somatic embryo with pCAMV was done with LBA4404(pCAMBIA2301). The constructs gave positive expression of GUS gene. The transient expression of GUS assay exhibited by cassava somatic embryo was blue with the constructs. There was increase in the number of somatic embryo that were stained blue with increase in the number of day of co-cultivation likewise co-cultivation of the explants in the light increases the number of explants that were stained blue. In the dark, 10% of the explants showed GUS gene expression after 2 days; after 4 days, the number of the explants increased to 40% while after 6 days 50% of the explants showed GUS gene expression. In the light, 30% of the explants showed GUS expression after 2 days; 50% showed GUS expression after 4 days and 65% expressed GUS gene after 6 days for the first cassava variety TME12 (Table 1). For the second cassava variety Albert, in the dark, 15% of the explants showed GUS gene expression after 2 days; after 4 days, the number of the explants expressed GUS gene increased to 40% while after 6 days 55% of the explants showed GUS gene expression. In the light, 25% of the explants showed GUS expression after 2 days; 50% of the explants expressed GUS gene after 4 days and 65% after 6 days (Table 2).

3.2 Expression of GUS Gene Using pOYE153

Genetic transformation using pOYE153 also showed expression of the GUS gene, the intensity using pCsVMV was 5 times deeper than the intensity expressed with pCAMV and as the number of the days of co-cultivation increases the colour intensity increase so therefore, the higher the number of day of co-cultivation the better the infection with *Agrobacterium*. The GUS gene expression in cassava somatic embryo increases as the number of days of co-cultivation increases, 6 days is better than 4 days and 4 days is better than 2 (Figs. 2 and 3). Samples that were co-cultivated in the light had higher GUS gene expression than the samples co-cultivated in the dark. The GUS staining expression was in parches for most of the explants. In the dark, 20% of the explants showed GUS gene expression after 2 days; after

4 days, this increased to 60% while after 6 days about 70% of the explants showed GUS gene expression. In the light, 40% of the explants showed GUS expression after 2 days; 75% after 4 days and 85% after 6 days for the first cassava variety TME12 (Table 1). For the second cassava variety Albert, In the dark, 20% of the explants showed GUS gene expression after 2 days; after 4 days, this increased to 55% while after 6 days about 70% of the explants showed GUS gene expression. In the light, 35% of the explants showed GUS expression after 2 days; 70% after 4 days and 90% after 6 days (Table 2). Overall, these experiments showed that both light and prolonged co-cultivation had a positive effect on gene expression in transient gene assays. 4% of the number of the explants used were 100% blue for the explants that were co-cultivated in the light.

The 35S promoter from CaMV is one of the most commonly used promoters in plant genetic transformation studies because it has the ability to drive high, constitutive gene expression in different tissues of the plant. In some *Agrobacterium* transformation vectors, the CaMV35S promoter is used to drive both the selectable marker and the gene of interest resulting in duplication of promoter sequences. This repetition of promoter sequences can cause gene silencing. Previous constructions were made either with the two CaMV 35S promoters following each other serially or with CaMV 35S facing each other [13]. This configuration causes overlapping transcription which can lead to transcriptional interference or gene silencing.

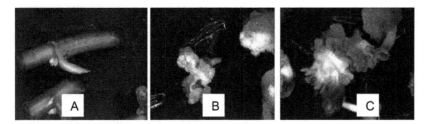

Fig.1. Preparation of cassava somatic embryo for genetic transformation
A- cassava nodal cuttings, B- induced somatic embryo, C- matured somatic embryo

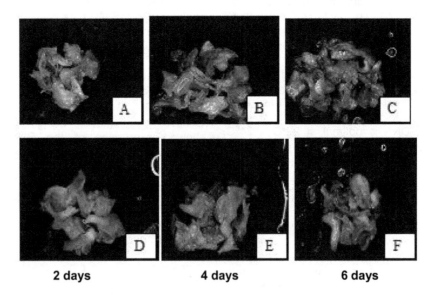

2 days   **4 days**   **6 days**

Fig. 2. Transient GUS gene expression of cassava somatic embryo in dark
A- GUS assay after 2 days on co-cultivation medium with Albert, B- GUS assay after 4 days on co-cultivation medium with Albert, C- GUS assay after 6 days on co-cultivation medium with Albert, D- GUS assay after 2 days on co-cultivation medium with TME 12, E- GUS assay after 4 days on co-cultivation medium with TME 12, F- GUS assay after 6 days on co-cultivation medium with TME 12

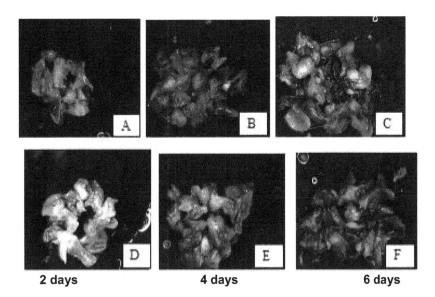

| 2 days | 4 days | 6 days |

Fig. 3. Transient GUSgene expression of cassava somatic embryo in light.
A- GUS assay after 2 days on co-cultivation medium with Albert, B- GUS assay after 4 days on co-cultivation medium with Albert, C- GUS assay after 6 days on co-cultivation medium with Albert, D- GUS assay after 2 days on co-cultivation medium with TME 12, E- GUS assay after 4 days on co-cultivation medium with TME 12, F- GUS assay after 6 days on co-cultivation medium with TME 12

Table 1. Percentage of the transformed explants for TME12

Number of days on co-cultivation	pCAMBIA2301 construct		pOYE153 construct	
	Dark (%)	Light (%)	Dark (%)	Light (%)
2	10	30	20	40
4	40	50	60	75
6	50	65	70	85

Table 2. Percentage of the transformed explants for Albert

Number of days on co-cultivation	pCAMBIA2301 construct		pOYE153 construct	
	Dark (%)	Light (%)	Dark (%)	Light (%)
2	15	25	20	35
4	40	50	55	70
6	55	65	70	90

This study revealed that pOYE153 with two different promoters: pCsVMV isolated from *Cassava Mosaic Virus* and pCaMV isolated from *Cauliflower Mosaic Virus*. The promoters pCsVMV designed in opposite directions; thereby transcription is in opposite direction further reducing the chances of gene silencing. The *Uid*A gene is used as the reporter gene in this project. The pCsVMV promoter has previously been tested by [14] and was found to be constitutive in tobacco and rice. Expression of GUS gene at transient level in cassava shows that pOYE153 was expressed at a higher level compared to the second construct pCAMBIA2301 with the GUS gene under control

of pCaMV. The experiment has shown that pCsVMV can be used to drive any other gene of interest. The project also showed that co-cultivation under light conditions aid genetic transformation and that longer periods of co-cultivation result in enhanced gene expression in transient assays.

4. CONCLUSION

This study has revealed that pCsVMV is a promoter that can drive the gene of interest just like the commonly used pCAMV. This also showed that the expression of GUS gene in explants can be improved by increasing the number of days in co-cultivation medium to six days as well doing the co-cultivation in the light.

COMPETING INTERESTS

Authors have declared that no competing interests exist.

REFERENCES

1. Buchanan BB, Gruissem W, Jones RL, eds. Biochemistry and molecular biology of plants. Rockville MD: American Society of Plant Physiologists. 2000;12:340-342.
2. Dale PJ, Clarke B, and Fontes EMG. Potential for the environmental impact of transgenic crops. Nature Biotechnol. 2002; 20:567-574.
3. Russell PJ, Genetics. 4th edn. New York: Harper Collins College; 1996.
4. Odell JT, Nagy F, Chua NH. Identification of DNA sequences required for activity of the cauliflower mosaic virus 35S promoter. Nature. 1985;3:13:810-812.
5. Matzke MA, Matzke AJM, and Mittelsten SO. Inactivation of gene-DNA-DNA interaction? In Homologous Recombination and Gene Silencing in Plants, The Netherlands: Kluwer Academic Publishers). 1994;271-307.
6. Assaad FF, Tucker KL, and Signer ER. Epigenetic repeat-induced gene silencing (RIGS) in Arabiopsis. Plant Mol. Biol. 1993;22:1067-1085.
7. Meyer P, Saedler H. Homology-dependent gene silencing in plants. Annual. Rev. Plant Physiol. Plant Mol. Biol. 1996;47:23-48.
8. Ingelbrecht IL., Mirkov TE, Dixon AG, and Menkir A. Epigenetic lessons from transgenic plants. Floriculture, Ornamental and Plant Biotechnology. 2006;2:88-97.
9. Opabode JT, Oyelakin OO, Akinyemiju AO, Ingelbrecht IL. Primary somatic embryos from axillary meristems and immature leaf lobes of selected African cassava varieties. British Biotechnology Journal. 2013;3:263-273.
10. Opabode JT, Oyelakin OO, Akinyemiju AO, Ingelbrecht IL. Influence of type and age of primary somatic embryo on secondary and cyclic somatic embryogenesis of cassava (*Manihot esculenta* Crantz). British Biotechnology Journal. 2014;4(3):254-269
11. Oyelakin OO, Opabode JT, Raji AA, Ingelbrecht IL. A cassava vein mosaic virus promoter cassette induces high and stable gene expression in clonally propagated transgenic cassava (*Manihot esculenta* Crantz). South African Journal of Botany. 2015;97:184-190.
12. Jefferson RA, Kavanagh TA, Bevan MW. Gus fusion: B-gluconidase as a sensitive and versatile gene fusion marker in higher plants. EMBOJ. 1987;6:3901-3907.
13. Benfey PN, and Chua NH. Regulated gene in transgenic plants. Science. 1989;244:174-181.
14. Verdaguer B, Kochko A, Beachy RN, and Fauquet, C. Isolation and expression in transgenic tobacco and rice plants of the cassava vein mosaic virus (CsVMV) promoter. Plant Mol. Biol. 1996;31:1129-1139.

11

Sequence Homology Studies of Phospholipase A$_2$-like Gene from Bloodstream form of *Trypanosoma brucei*

Ishaya Y. Longdet[1*], Hajiya M. Inuwa[2], Isma'il A. Umar[2] and Andrew J. Nok[2]

[1]*Department of Biochemistry, University of Jos, Nigeria.*
[2]*Department of Biochemistry, Ahmadu Bello University, Zaria, Nigeria.*

Authors' contributions

This work was carried out in collaboration between all authors. Author AJN designed the study and wrote the protocol. Author IYL managed the analyses of the study and wrote the first draft of the manuscript. Authors HMI and IAU performed the Bioinformatics analysis and the literature searches. All authors read and approved the final manuscript.

Editor(s):
(1) Laura Pastorino, Dept. Informatics, Bioengineering, Robotics and Systems Engineering (DIBRIS), University of Genoa, Italy.
(2) Chung-Jen Chiang, Department of medical laboratory Science and Biotechnology, China Medical University, Taiwan.
Reviewers:
(1) Anonymous, University of New Mexico and New Mexico VA Health Care System, Albuquerque, USA.
(2) Anonymous, Federal University of Rio de Janeiro, Brazil.
(3) Anonymous, Universidad Veracruzana, México.
(4) Anonymous, University Brunei Darussalam, Brunei.
(5) Anonymous, Simon Bolivar University, Venezuela.

ABSTRACT

Aim: This work focused on the sequence homology studies of the enzyme, phospholipase A2 (PLA$_2$), in *Trypanosoma brucei* obtained from the blood of bull in Federe, Plateau State, Nigeria, West Africa

Place and Duration of Study: Department of Biochemistry, University of Jos, Nigeria; Department of Biochemistry, Ahmadu Bello University, Zaria, Nigeria, Department of Biotechnology, NVRI, Vom, Nigeria; between June 2009 and September 2011.

Methodology: *T. brucei* grown in rats were harvested and separated using diethyl amino ethyl (DEAE) cellulose chromatography. From the parasites' genomic DNA the PLA$_2$-like gene was amplified using consensus primers. The amplicon was cloned unto pMal-2cE vector and confirmed using direct PCR and restriction enzyme analyses. The PLA$_2$ gene and translated protein

Corresponding author: Email: islongdet@yahoo.com

sequences were studied using National Center for Biotechnology Information (NCBI) Conserved Domain Search Tool and Conserved Domain Architectural Retrieval Tool

Results: Analyses of the 1344bp gene sequence using bioinformatics tools showed that it is very closely related to PLA_2 sequences of *T. brucei* (TREU 927) and *T. b. gambiense*. Motifs that are unique to PLA_2 (FSHGL) and lipases (GHSFG) were found to be present in the query sequence. The domains present in the studied sequence agreed closely with those of the human platelet activating factor acetyl hydrolase (PAF-AH). There was also a good sequence resemblance with PLA_2s from *T. cruzi*, *Metarhizium amisop*, *Metarphizium acridu* and PAF-AH in terms of architecture.

Conclusion: The PLA_2-like gene isolated from the blood stream form of *Trypanosoma brucei* and studied was found to posses the domains and motifs unique to PLA_2s and lipases and so homology was established among the proteins.

Keywords: Trypanosoma brucei; Phospholipase A_2; gene; motif; domain.

1. INTRODUCTION

Phospholipase A_2 (3.1.1.4) comprises of a diverse family of enzymes that hydrolyze glycerophoshosphospholipids at the sn-2 position giving rise to free fatty acids and lysophospholipids. The membership of the PLA_2 super family is ever expanding [1]. The enzymes have been classed into 15 groups [2]. These groups were regrouped into two – those utilizing a catalytic histidine and those using a catalytic serine. All the members of the PLA_2 super family carry a consensus sequence GXSXG which is common to many other lipases [3] despite some differences that are the basis for the various patterns of classification. The enzymes are classed based on their source, amino acid sequence, chain length and disulphide bond patterns [4] as well as based on the guiding information that the enzyme must catalyze the hydrolysis of the sn-2 ester bond of a natural phospholipid substrate. In addition, the enzyme must have complete protein sequence of the mature protein established; have homologous enzymes distinguished and the spliced variant established within subgroups; have the sequence homology and catalytic activity established in order to be classed along with others [1].

The physiological role of PLA_2 gene expressed as PLA_2 protein includes the hydrolysis of phospholipids yielding free fatty acids (arachidonic acid and oleic acid) and lysophospholipid. The fatty acids are important stores of energy. Arachidonic acid is an important metabolic intermediate for producing eicosanoids, which are regulatory factors implicated in a wide range of physiological and pathological states serve as potent mediators of inflammation and signal transduction [5]. The other product, lysophospholipid, is important in cell signaling, phospholipid remodeling and membrane perturbation [3]. This enzyme is also reported to regulate the entry of calcium ions into *Trypanosoma brucei* and help the parasite modulate the host-parasite interaction [6] thereby implicating PLA_2 in the pathogenesis of the *T. brucei* The PLA_2 family has become a major drug target for many different diseases [7,8].

PLA_2 has been reported in different species of trypanosomes to various degrees. Some reports show that phospholipase A_2 has been identified to have orthologs in *T. brucei*, *T. cruzi*, and humans and the *Leishmania* PLA_2/PAF-AH (LmjF.35.3020) contains a predicted N-terminal signal peptide sequence and transmembrane domain and a predicted lipase/platelet-activating factor acetylhydrolase sequence [9]. This group of enzymes has been isolated from various sources such as animal toxins like snake venom [10]; insect venom [9] and mammalian organs [11,12]. Some other reports show that the enzyme has also been isolated and purified to electrophoretic homogeneity in *T. congolense* [13]; that activity and kinetics of PLA_2 from *T. brucei gambiense* and *T. brucei* have been detected and studied [14]. The *Trypanosoma brucei* (Tb09.211.3650) Gene Data Base on TritrypDB hosted by the Sanger Institute, revealed a protein encoding gene was present having a product described as phospholipase A_2-like protein, putative. However, reports on PLA_2 in *T. brucei* distributed around the endemic regions are quite scarce. Equally, the PLA_2 from *Trypanosoma species* has not found inclusion in the several classes of the enzymes reported in literature possibly due to the scarcity of literature on its studies. Therefore, this work was designed to study the gene sequence of PLA_2 obtained from *T. brucei* in Nigeria, West Africa, for the first time, in comparison with those already characterized. This may contribute in the understanding of the nature of the enzyme and so in subsequent classification. As part of the effort, Phospholipase A$_2$-like gene from blood stream form *Trypanosoma brucei brucei* had been reported to conserve domains and motifs peculiar to characterized PLA$_2$s and lipases [15]. Literatures have revealed that genome sequence analysis have helped in the rcharacterization of a PLA$_2$ cDNA from *Arabidopsis thaliana* [16] genes cncoding hctcrodimeric phospholipases A$_2$ from the scorpion *Anuroctonus phaiodactylushav* [17] sequences and structural organization of phospholipase A$_2$ genes from *Vipera aspis aspis*, *V. aspis zinnikeri* and *Vipera berus berus* venom [18] plasmid pLA1 present in *N. pentaromativorans* US6-1 [19]; this approach has therefore found application in Metagenomics [20].

2. MATERIALS AND METHODS

2.1 Reagents and Equipment

Chemicals used were of analytical grade and purchased from Sigma and Pharmacia Fine chemicals. DNA extraction kit was purchased from Bio Basic Inc. Markham Ontario, Canada. *Taq* polymerase and High Fidelity Polymerase Enzyme Mix were bought from Promega, USA and Fermentas, respectively. High Pure polymerase chain reaction (PCR) Clean-Up Kit used to purify PCR products was purchased from Fermentas and 100 bp DNA molecular size marker was purchased from Roche, Mannheim Germany. Products from Fermentas and oligonucleotide primers were supplied and synthesized by Inqaba biotec Industry®, Pretoria South Africa. GeneAmp PCR System9700 used for amplification was obtained from Applied Biosystems, Indonesia and the Gel Documentation System was obtained from Synegene® Inc. Indonesia. Sequencing analyses were performed by Inqaba Biotec Industries®, Pretoria South Africa.

2.2 Parasites Isolation

An isolate of *T. b. brucei* from cows (Federe isolates) were obtained from the Parasitology Department, NITR, Vom. Adult albino rats were infected through intra peritoneal route with 0.2ml *Trypanosoma brucei* infected blood diluted with 2ml phosphate saline glucose (PSG) buffer pH 8.0 to give cell density of $\approx 1 \times 10^6$ cells/ml. Parasitemia was monitored by wet smear via tail snip. At peak parasitemia ($\approx 1 \times 10^8$ cells/ml),

the rats were euthanized and blood collected. Parasites were purified on DEAE-cellulose (pre-swollen whatman DE-52-Pharmacia Fine Chemicals) as previously described [21].

2.3 DNA Isolation

Genomic DNA was extracted from 200µl ($\approx 1 \times 10^8$ cells/ml) of isolated *T.b. brucei* suspended in PSG buffer (0.6g NaH$_2$PO$_4$; 0.71g Na$_2$HPO$_4$ and 14.61g NaCl in 450ml distilled water adjusted to pH 8.0 with orthophosphoric acid) using ZR Genomic DNA Tissue Minipreps Kit (Zymo Research) according to the manufacturer's instructions. Briefly, cell lysis was with 500 µl lysis buffer (which consisted of 10 mM phosphate buffer containing and protease inhibitor) and proteinase K (3 µl of 10mg/ml) incubated at 55°C for 30 minutes. DNA was precipitated with 260 µl absolute ethanol. Precipitated DNA was captured in EZ-10 column by centrifuging at 12,000 x g for 1minute. The flow through was discarded while the EZ-10 column placed in a fresh vial and centrifuged again at 12,000 x g for 1minute to remove residual wash buffer. DNA was eluted into 1.5 ml microcentrifuge tubes with 50 µl elution buffer after incubation at 50°C for 2 minutes by centrifuging at 14,000 x g for 1 minute.

2.4 PCR Amplification of Phospholipase A$_2$ Like Gene

Phospholipase A$_2$ like gene was detected and amplified by Polymerase Chain Reaction (PCR) with primers designed based on the gene sequence of PLA$_2$ like gene in the GeneBank Data Base (Tb 09.211.3650, Phospholipase A$_2$-like protein, putative, *T. brucei*, chr 9). The primer designed was done in Inqaba Biotech Industry, Pretoria, South Africa. The primers used were as follows: Sense primer 5'-ATGGTAACGTGGGC GCTGAA GTAT- 3'carrying *BamHI* site and Anti-sense primer 5'-CTAACACGTTGAACACACTTC GGTA-3'carrying *PstI* site. High Fidelity Taq DNA Polymerase Enzyme kit (Fermentas) was used to amplify the gene from the genomic DNA according to the manufacturer's instructions. The optimum reaction mix in 50 µl volume was as follows: nuclease free water (37.6µl); 10 x PCR buffer (5.0 µl); dNTP mix (1.0 µl); each primer (1.0 µl); High Fidelity Taq enzyme mix (0.4 µl); Genomic DNA (5.0 µl of 50µg DNA/ml). The thermal cycling was carried out with the following process profile: initial denaturation at 94°C for 2minutes, elongation 94°C for 30 seconds, 56°C for 30 seconds, 68°C for 2 minutes running for 30 cycles, and final extension at 68°C for 10 minutes; then ending/waiting at 4°C for ∞. Ten microliters (10 µl) of the product was separated on 1.0% agarose gel to check the success of the process and the results documented using Gel Documentation System (Synegene®).

2.5 Cloning and Sequencing

The PLA$_2$ gene amplified by PCR from *T. b. brucei* was purified using High Pure PCR Product Clean-Up Kit (Fermentas) and ligated into *pMal-c2E* vector in a 20 µl ligation reaction. The ligation reaction mixture was incubated at 16°C over night and subsequently used for transformation. The recombinant plasmid was designated *pMal-PLA$_2$* and transformed into *Escherichia coli* DH5α competent cells. Transformants were placed on LB agar containing 100µg/ml ampicillin. PCR Cloning Kit (Fermentas) was used to clone the purified PLA$_2$ from *T. b. brucei*. Colonies were randomly selected for screening of positive clones by PCR and restriction endonucleases digestion of plasmids using *BamHI* and *Pst I*. The sequence was submitted to the Gen Bank Data Base.

2.6 Transformation of Clone into *E. coli* (BL-21 (DE3)) and Expression

One micro liter of the ligated DNA construct was transformed into *E. coli* expression grade BL-21 (DE3) competent cells (Lucigen) according to the manufacturers' instruction. The colonies of the BL-21 (DE3) cells were picked from the agar plates and inoculated in 30ml broth medium (SOC plus 100µl/ml ampicillin) and incubated for 3 hours at 37°C. The cells were then harvested by centrifugation at 5000g for 15 minutes at 4°C. The medium was discarded from control and induced cells. The cell pellets were re-suspended in 5 ml lysis buffer and sonicated for short pulses of 10 seconds for 4 times on ice. The lysates were centrifuged at 10, 000g for 20 minutes at 4°C. The supernatant which had the protein was collected. The pellet was re-suspended in 5 ml lysis buffer. All supernatants were then analyzed on 10 % SDS-PAGE.

2.7 Bioinformatics Sequence Analysis

The Finch TV® programmes (GeoPiza) was used to analyze the PLA$_2$ gene while NCBI BLAST programmes such as NCBI Conserved Domain Search Tool (CDD) and NCBI

Conserved Domain Architectural Retrieval Tool (CDART) were used to study the PLA$_2$ gene and translated protein sequences in order to establish the possibility of homology or similarity.

3. RESULTS

The PCR amplification gave a 1.3kb PLA$_2$ like band from the gene from genomic DNA of *Trypanosoma brucei*. The amplicon cloned unto pMal-c2E vector between the restriction sites of *BamH1* and *Pst1* yielded pMal-PLA$_2$ of about 7.9kb which is approximately the size of the pMal-c2E vector (6.6kb) and PLA$_2$ (1.3kb) put together. Direct PCR analysis done to confirm the quality of the insert DNA gave amplicon of about 1.3kb which was similar in size with the PLA$_2$ like gene amplified from the genome DNA of bloodstream form *T. b. brucei*. The digestion of the purified pMal-PLA$_2$ clone and separation on agarose gel electrophoresis equally revealed the pMal-2cE vector with about 6.6kb as well as the PLA$_2$ like gene (1.3kb). These confirmed the cloning of the PLA$_2$ like gene into pMal-2cE to form pMal-PLA$_2$ clone. The heterologous expression of *T. brucei* PLA$_2$ in pMal-c2E plasmid recorded a success in the transformation process of the competent *E. coli* cells. On the other hand, the heterologous expression of the recombinant PLA$_2$ in BL-21 (DE3) competent *E. coli* cells was not successful. The corresponding fractions obtained from the colonies were resolved on SDS-PAGE. This put a limit to biochemical characterization of protein in order to confirm the finding from bioinformatics analysis.

The gene sequence with 1, 344bp nucleotides primary sequence obtained has been assigned the accession number *JN603736*. The BLASTN Alignment view programme compared the PLA$_2$ like gene sequence with gene sequences were from *T. brucei* TREU927 PLA$_2$ like protein (Gene ID: Tb 3661014 Tb 09.211, 3650; XM 822413.1) and *T. brucei gambiense* (DAL 972, FN 554972.1).

The detail NCBI BLASTN alignment view (Fig. 1) showed that no deletion occurred in the PLA$_2$ like gene (query) except eight substitutions distributed at positions 51, 61, 178, 282, 530, 810, 888 and 970 as compared with the two gene sequences in the Gene Bank DB. The substitutions in positions 51, 61, 282, 530, and 888 do not change the amino acid in those respective positions because the new codons still coded for the same amino acids. On the other hand, the substitutions in positions 178, 810 and

970 caused changes in amino acids in those positions i.e. at position 78 [TTG changed to TTC]: TTG codes for Leucine while TTC codes for Phenylalanine; at position 810 [AGT changed to ATT]: AGT codes for Serine while ATT codes Isoleucine and at position 970 [GTA changed to CTA]: GTA codes Valine while CTA codes for Leucine.

The parasite Genomes WU-BLAST2 analysis of the sequence gave 99% identity and similarity to the *T. brucei* (TREU 927) PLA$_2$ like sequence from partial mRNA, chromosome 9 and *T. brucei gambiense* DAL 972 Chr 9, complete sequence PLA$_2$ (Table 1). The result also revealed that there was 100% coverage, no gaps, 99% identity and zero E – values in each comparison.

The translated PLA$_2$ primary structure of 447 amino acids was compared with GeneBank PLA$_2$ sequence Databases in NCBI using blastx established homology with those of other putative PLA$_2$ proteins from *T. brucei* (TREU927), *T. b gambiense* (DAL972), *T. cruzi*, *Leishmannia major* and characterized human Platelet – Activating Factor Acetyl Hydrolase (Fig. 2). The alignment showed that the protein carries a conserved *GHSFG* lipase motif and an *FSHCL* motif peculiar to PLA$_2$s which are also conserved in the query. These significantly elaborated the query sequence similarity with the PLA$_2$ family.

Query	1	ATGGTAACGTGGGCGCTGAAGTATTTTGTTCGCGTAGTCCGATGGTCGACAGAAGCATTC	60
TREU927:	61	....A.........	60
DAL972:1656744		....G.........	1656803
Query	61	CTAATTTGGCCCACACGGCCACTTTTTGACTATGCCACCTCATTGCATTGTGTTCCCATA	120
TREU927	61	C..	120
DAL972:1656804		T..	1656863
Query	121	AGCGGCACATTTATTACTTCGGTTCTGCTCTGCTTCTACGGGTTCCCACTTTTGTGGTCT	180
TREU927	121	....C......	180
DAL972:1656864		....G......	1656923
Query	241	CCACTATTGAAACCCATTGGCGGTCGCTATAGCGTGGGCCTCGTGCATATGAACGGCTGC	300
TREU927	241	...........................T................	300
DAL972:1656984		...........................C................	1657043
Query	481	AACGCGGTGCCCGCCGCTTTGCTCAACCAATATGAGAGAGTGCCGCCTATTGTTGTGTTC	540
TREU927	481	....A.............	540
DAL972:1657224		....G.............	1657283
Query	781	AAGGACTTTTGGACAACTTTGGGCTACAGTAATTCAGATATTGACAAGTTTCTTAGCAAA	840
TREU927	781	..........................T........................	840
DAL972:1657524		..........................G........................	1657583
Query	841	CCGTTGCAGGTACATCTTGCGGGTCATTCATTTGGCGGTGCCACTGTACTCGCGGCTGCA	900
TREU927	841	:::A:::::::::::	900
DAL972:1657584		....G.............	1657643
Query	961	CCATGCATGGTACCAATACAAAATGAACATTTTTGCAACCCGCTTTCTGATGCCCGTAAA	1020
TREU927	961	.........G..	1020
DAL972:1657704		....	1657763
Query	1321	ACCGAAGTGTGTTCAACGTGTTAG	1344
TREU927	1321	ACCGAAGTGTGTTCAACGTGTTAG	1344
DAL972:1658064		ACCGAAGTGTGTTCAACGTGTTAG	1658087

Fig. 1. NCBI BLASTN Alignment View of PLA$_2$ like gene sequence with *T. brucei* TREU927 PLA$_2$ like protein from partial mRNA length = 1344(Gene ID: Tb 3661014 Tb09.211.3650; XM 822413.1) and *T.brucei gambiense* DAL972, chromosome 9, complete sequence length = 2,160,261 (FN554972.1)

Table 1. Comparison of PLA$_2$-like gene sequence with those of two sequences in data base (DB) using NCBI and Parasite Genomes Washington University Basic Local Alignment Search Tool 2 (WU-BLAST2)

DB- AN	Source	Max score	Total score	Length(bp)	Query coverage (%)	Identity (%)	Gaps (%)	E.value
FN554972.1	*T. brucei gambiense*	2640	2640	2,160,261	100	99	0	0
XM841212.1	*T.brucei TREU927*	2640	2640	1,344	100	99	0	0

Each source parasite is identified by its data base accession number (DB-AN) and the sequence length given in base pairs (bp)

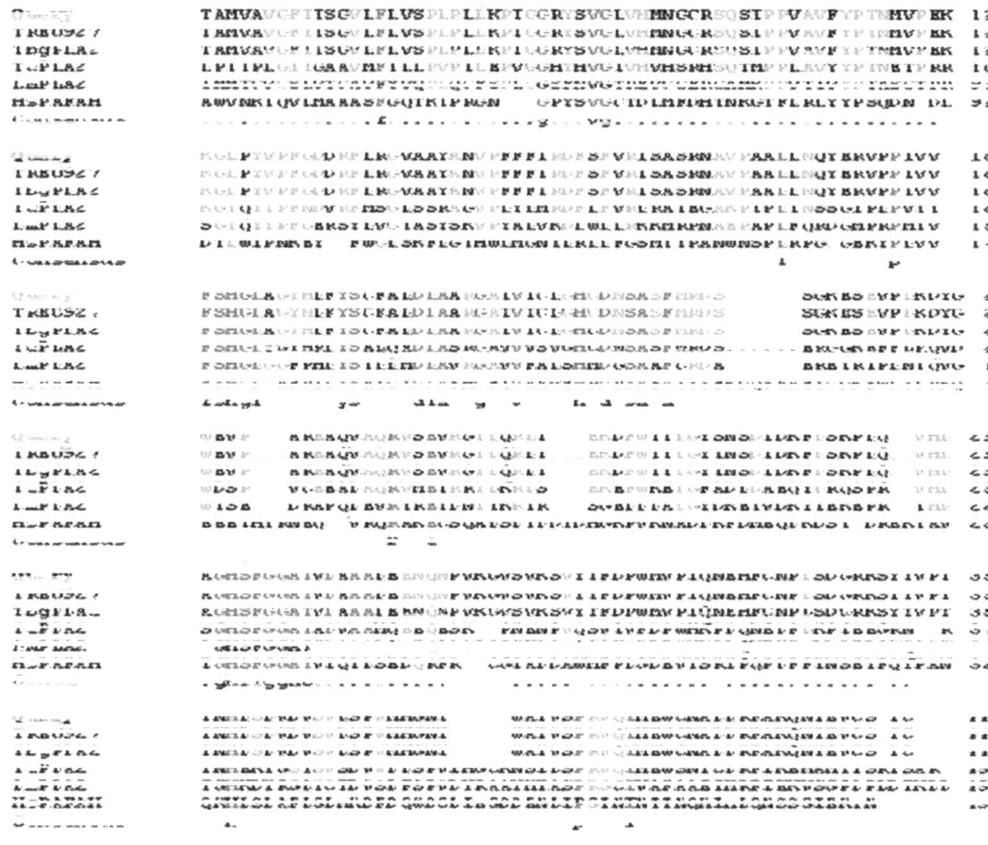

Fig. 2. Amino acid sequence alignment of PLA$_2$-like protein with PLA$_2$ from *T. brucei* (TRUE927), *T. b. gambiense* (TbgPLA$_2$), *T. congolense* (TcPLA$_2$), *Leishmania major* (LmPLA$_2$) and PAF-AH. Conserved motifs and lipase consensus motif were marked with red colour

The NCBI conserved Domain search Tool analysis of the protein sequence predicted the Conserved domains (Fig. 3). The result showed that the PLA$_2$ translated protein sequence which had 447 amino acid residues was mapped to PAF-AH super family, a sub-family of PLA$_2$ super family. The conserved domains were marked in red ink. Equally predicted was the Conserved Domain Architecture of the protein sequence using the Conserved Domain Architecture Retrieval Tool (CDART).

The results (Fig. 4) showed the graphic view of conserved domains architecture on PLA$_2$ like protein. Proteins with similar architectures were PLA$_2$ from *T. cruzi*, *Metarhizium amisop*, *Metarphizium acridu* and PAF-AH.

4. DISCUSSION

The gene sequence (1,344 bp) presented and analyzed using some Bioinformatics Tools revealed some interesting high lights. The revealed 99% identity and similarity between the PLA$_2$ – like gene from *T. b. brucei* and that of *T. brucei* (TREU 927) PLA$_2$ – like sequence from partial mRNA, chromosome 9 and *T. b gambiense* (DAL 972 chromosome 9, PLA$_2$ complete sequence) as shown (Table 1) signify that they are homologous. Also the translated protein sequence of the PLA$_2$-like gene from *T. b. brucei* (Fig. 2) revealed identities and similarities in the conserved domains region with members of the Platelet Activity Factor Acetyl Hydrolase (PAF-AH) super family. These findings agree with previous reports [9]. This further substantiated by the report that PAF-AH is a subfamily of the PLA$_2$ super family responsible for inactivation of platelet-activating factor through the cleavage of an acetyl group [22,23]. The conserved domains among the protein sequences were presented (Fig. 3) revealing that large portions of the sequence were conserved. The sequential order of the conserved domains in the protein sequence was also conserved as revealed by the CDART analysis. The sequence of PLA$_2$ like protein from *T. b. brucei* was similar in architecture (Fig. 4) to the sequences of PLA$_2$ from *T. cruzi*, *M. anisop*, *M. acridu* and PAF-AH in the Genebank DB. This implies that the sequences are similar in architecture and not just ordinary sequence similarity. Since the conserved Domain Database brings together several collections of multiple sequence

alignment with conserved domains [22] and the CDART performs similarity searches based on the sequential order of the conserved domains, get them grouped and scored by architecture [21], then the PLA$_2$ like gene (JN603736) used in this study was a PLA$_2$ gene of the *T. b. brucei*. Other important features conserved in the studied sequence were the lipase motif (*GHSHG*) and the motif (*FSHGL*) peculiar to PLA$_2$s. The sequences of PLA$_2$ from *T. brucei* (TREU 927), *T. b. gambiense*, *T. cruzi*, *Leishmania major* and characterized PAF-AH shared consensus sequences along with the conserved motifs with the query protein as produced by the NCBI BLASTX sequence alignment. This further elaborated the sequence similarity of the PLA$_2$ like protein from *T. b. brucei* with the PLA$_2$ super family as a justification that it was a PLA$_2$ gene that was amplified and studied. This is another case of homology which had been reported to be common among the members of the PLA$_2$ super family [1,24]. These conserved domains and motifs are similar to those reported in a study using *T. brucei* (strain EATRO 427 clone MITat 1.2) [25]. Also, the use of Trans-sialidase-like gene from the bloodstream form of *Trypanosoma evansi* that conserved most of the active siteresidues and motifs found in *Trypanosomal sialidases* and trans-sialidases has been reported as a basis for homology [26].

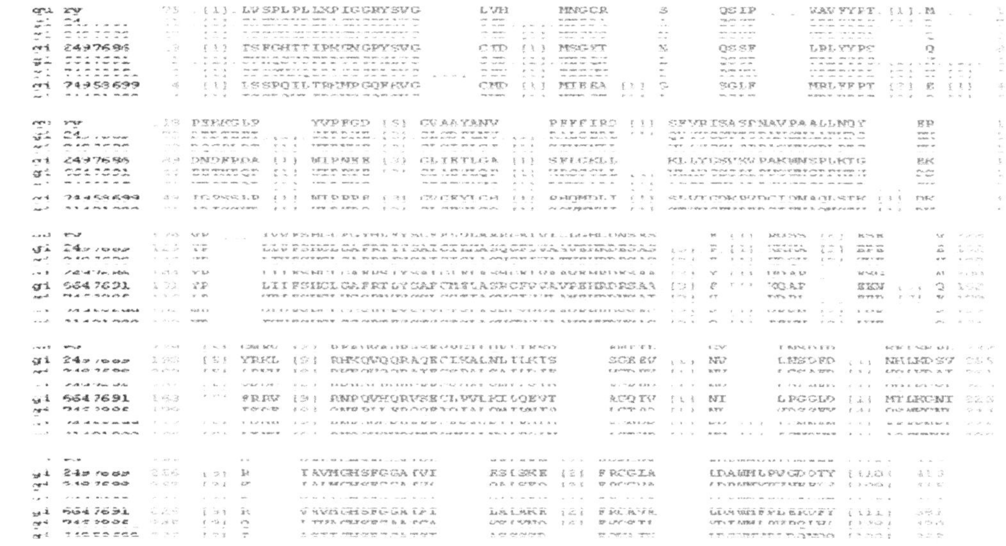

Fig. 3 Conserved Domains in PLA$_2$-like protein as aligned to members of PAF-AH super family. This was shown in red colour signifying high relatedness while domains that were weakly related were indicated in blue colour. Unaligned portions were left and indicated by numbers in brackets. The members of the PAF-AH super family used in the alignment were represented by their respective identification numbers

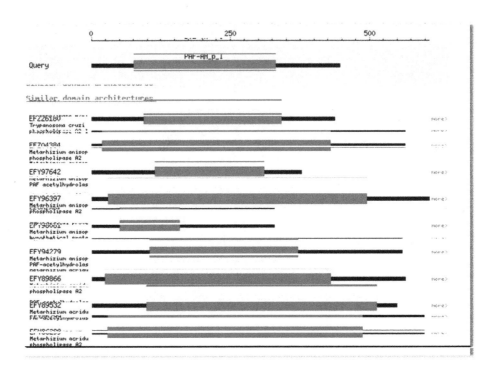

Fig. 4. Conserved domains architecture on PLA₂ like protein. Proteins with similar architectures include PLA₂ from *T.cruzi*, *Metarhizium anisop*, *Metarhizium acridu* and PAF-AH

The Alignment view (Fig. 1) revealed only substitutions in the PLA₂ like gene at positions 178, 810 and 970 that caused changes in the amino acids at the respective positions. These alterations did not appear in the active center (GHSFG) of the enzyme and so the activity of this enzyme may remain unaltered. This revelation agrees with conclusions that the nucleotide substitution can change the triplet sequence (codons) and hence can cause redundancy of the protein especially when it occurs on the catalytic domain [27]. However, the changes in such protein sequences may also occur at positions irrelevant for enzymatic activity [28]. These findings may provide ground for considering the enzyme a possible member of one of the classes in the superfamily.

5. CONCLUSION

Conclusively, the presence of PLA₂ and lipase motifs, PLA₂ conserved domains and the high percentage identity and similarity of the PLA₂ - like sequence to some characterized PLA₂ primary sequences indicate that it is a PLA₂ homologue of the Platelet-activating factor acetylhydrolase superfamily. The unsuccessful expression of the recombinant PLA₂ in BL-21 (DE3) competent *E. coli* cells as displayed on fractions obtained from the colonies resolved on SDS-PAGE suggests that alternative ways for the expression need to sought in order to biochemically characterize the gene product.

ETHICAL APPROVAL

Animal experiments were carried out in accordance with the instructions for the care and use provided by the university of Jos, Nigeria where the animal experiments were carried out. The experiments were examined and approved by the university of Jos ethics committee

ACKNOWLEDGEMENTS

We acknowledge the Tertiary Education Trust Fund (TETFUND), Abuja, Nigeria for supporting this work with grant. We appreciate the technical assistance of the staff of the Molecular Biology Laboratory, NVRI, Vom, Nigeria as well as the kindness of the staff of the Parasitology Department, NITR, Vom, Nigeria for supplying the parasites.

COMPETING INTERESTS

Authors have declared that no competing interests exist.

REFERENCES

1. Six DA, Dennis EA. The expanding superfamily of phosphoplipase A_2 enzymes: Classification and characterization. Biochem. Biophy. Acta. 2000;1488:1-19.
2. Schaloske RH, Dennis EA. The phospholipase A_2 superfamily and its group numbering system. Biochem. Blophy. Acta. 2006;1761:1246-59.
3. Balsinde J, Balbao MA, Insel PA, Dennis EA. Regulation and inhibition of phospholipase A_2. Annu. Rev. Pharmacol. Toxicol. 1999;39:175-89.
4. Modesto, JC, Spencer PJ, Fritzen M, Valenca RC, Oliva MLV, de Silva MB, et al. BE-I-PLA_2, a novel acidic phospholipase A_2 from Bothrops erythromelas venom: Isolation, cloning and characterization as potent anti-platelet and inductor of prostaglandin I_2 release by endothelial cells. Biochem. Pharm. 2006;72:377-84.
5. Vadas P, Pruzanski W. Role of Secretory PLA_2 in the pathobiology of disease. Laboratory Investigation. 1986;55:391–398.
6. Eintracht J, Maathai R, Mellors A Ruben L. Calcium entry in Trypanosoma brucei is regulated by phospholipase A_2 and arachidonic acid. The Biochemistry Journal. 1998;336:659-666.
7. Gururao H, Manoj K, Punit K, Tej PS, Ramaswamy PK. Human group III PLA_2 as a drug target: Structural analysis and inhibitor binding studies. International Journal Of Biological Macromolecules. 2010;47(4):496-501.
8. Burke JE, Dennis EA. Phospholipase A_2 structure/function, mechanism, and signaling. Journal of Lipid Research. 2009;237-242.
9. Pawlowic M. Phospholipid Metabolism of Leishmania Parasites. A PhD Dissertation In Biology. Texas Tech University, USA; 2013.
10. Abubakar MS, Nok AJ, Abdurahman AK, Haruna AK, Shok M. Purification of two Phospholipase enzymes from Naja n. nigricolis Reinhardt venom. J. Biochem. Mol. Toxicol. 2003;17(1):53-58.
11. Prenner C, Mach L, Glossl J, Marz L. The antigenicity of the carbohydrate moiety of an insect glycoprotein, honey-bee (Apis mellifera) venom phospholipase A_2: The role of a 1,3-fucosylation of the asparagine-bound N-acetylglucosamine. Biochem. J. 1992;284:377-380.
12. Valentin E, Ghomashchi F, Gelb MH, Lasdunski M, Lambeau G. On the diversity of Secreted PLA_2s. J. Biol. Chem. 1999;274(29):31195-31202.
13. Nok AJ, Esievo KAN, Ibrahim S, Ukoha AI, Ikediobi CO. Phospholipase A_2 from Trypanosoma congolense: Charaterization and haematological properties. Cell Biochem. Function. 1993;11:125-130.
14. Shuaibu MN, Kanbara H, Yangai T, Ameh DA, Bonire JJ, Nok AJ. PLA_2 from T.brucei gambiense and T. b. brucei: inhibition by organotis. J. Enzyme Inhib. 2001;16:433-441.
15. Longdet IY. Phospholipase A_2-like gene from blood stream form Trypanosoma brucei brucei conserves domains and motifs peculiar to characterized PLA_2s and lipases. 3rd International Conference on Proteomics and Bioinformatics (OMICS Group Conferences) July 15 – 17, 2013; Philadelphia, USA: http://omicsgroup.com/conferences/proteomics-bioinformatics-2013/
16. Ryu SB, Leeb HY, Doellinga JH, Paltaa JP. Characterization of a cDNA encoding Arabidopsis secretory phospholipase A_2: an enzyme that generates bioactive lysophospholipids and free fatty acids. Biochim. Biophys. Acta. 2005;1736:144 – 151.
17. Valdez-Cruz NA, Segovia L, Corona M. Sequence analysis and phylogenetic relationship of genes encoding heterodimeric phospholipases A_2 from the venom of the scorpion Anuroctonus phaiodactylus. Gene. 2007;396(1):149-58.
18. Guillemin I, Bouchier C, Garrigues T, Wisner A, Choumet V. Sequences and structural organization of phospholipase A_2 genes from Vipera aspis aspis, V. aspis zinnikeri and Vipera berus berus venom: Identification of the origin of a new viper population based on ammodytin I1 heterogeneity. Euro. J. Biochem. 2003;270(13):2697–2706.
19. Yun SH, Choi C-W, Lee S-Y, Lee YG, Kwon J, et al. Proteomic Characterization of Plasmid pLA1 for Biodegradation of Polycyclic Aromatic Hydrocarbons in the Marine Bacterium, Novosphingobium pentaromativorans US6-1. PLoS ONE. 2014;9(3):e90812. DOI:10.1371/journal.pone.0090812.

20. Wood DE, Salzberg SL. Method Kraken: Ultrafast metagenomic sequence classification using exact alignments. Genome Biology. 2014;15:R46. DOI:10.1186/gb-2014-15-3-r46

21. Lanham SM, Godfrey DG. Isolation of Normal plasma lactose concentrations and kinetics of salivarian trypanosomes from man and mammals intravenously infused lactose in cattle: Research in using DEAE cellulose. Experimental Parasitology and Veterinary Science. 1970;65:1-4.

22. Derewenda ZS, Derewenda U. The Structure and Function of PAF-AH. Cell Mol. Life Sci. 1998;54(5):446 – 55.

23. Marchler-Bauer A, Lu S, Anderson JB, Chitsaz F, Derbyshire MK, DeWeese-Scott C, et al. CDD: a conserved domain database for the functional 207 annotation of proteins. Nuc. Acids Res. 2011;39:225-9.

24. Geer LY, Domrachev M, Lipman DJ, Bryant SH. CDART: Protein homology by domain architecture. Genome Res. 2002;12(10):1619-23.

25. Muhammad K. Identification and characterization of phospholipase A_2 from *Trypanosoma brucei. A PhD* Dissertation. Faculty of Chemistry and Pharmacy of the University of Tübingen; 2009.

26. Yakubu B, Nok AJ, Sanni I, Inuwa HM. Trans-sialidase-like gene from the bloodstream form of *Trypanosoma evansi* conserves most of the active site residues and motifs found in *Trypanosomal sialidases* and trans-sialidases. African J. Biotech. 2011;10(13):2388-2398.

27. Granner DK. Protein Synthesis and the Genetic Code. In: Murray RK, Granner DK, Mayes PA, Rodwell VW, editors. Herper's Illustrated Biochemisstry 26th Edition, USA: McGraw-Hill. 2003;368 -73.

28. Motagna G, Cremona LM, Paris G, Amaya FM, Bushiazzo A, Alzari, M, Frasch CCA.. Transialidase from the African trypanosome, *Trypanosoma brucei.* Eur. J. Biochem. 2002;269:2941–50.

12

Assesment of Cocoa Genotypes for Quantitative Pod Traits

Olalekan Ibrahim Sobowale[1*], Benjamin Oluwole Akinyele[2]
and Daniel Babasola Adewale[3]

[1]*Crop Improvement Division, Cocoa Research Institute of Nigeria, Ibadan, Nigeria.*
[2]*Department of Crop, Soil and Pest Management, The Federal University of Technology Akure,
Ondo State, Nigeria.*
[3]*Department of Biological Sciences, Ondo State University of Science and Technology, Okitipipa,
Ondo State, Nigeria.*

Authors' contributions

*This work was carried out in collaboration between all authors. Author OIS designed the study,
performed the statistical analysis, wrote the protocol, and wrote the first draft of the manuscript.
Authors BOA and DBA managed the analyses of the study. Author OIS managed the literature
searches. All authors read and approved the final manuscript.*

Editor(s):
(1) Yan Juan, Doctorate of Horticultural Crop Biotechnology Breeding, Sichuan Agricultural University, Ya'an, China.
(2) Kuo-Kau Lee, Department of Aquaculture, National Taiwan Ocean University, Taiwan.
Reviewers:
(1) Anonymous, Brazil.
(2) Anonymous, South Africa.
(3) Anonymous, China.

ABSTRACT

Yield related traits in 20 cocoa genotypes were investigated to determine suitable parental
genotypes for yield improvement programmes in cocoa. Fifteen uniformly ripe pods were collected
for pod and bean characteristic assessment from twenty genotypes in an existing cocoa hybrid trial
research plot laid out in a randomized complete block design with six replications at the Cocoa
Research Institute of Nigeria (CRIN), Ibadan, Nigeria. Seven quantitative data on the pods were
subjected to statistical analysis. The 20 genotypes differed significantly ($P < 0.001$) for the seven
traits. Performances of the genotypes ranged as: pod weight (175.40 – 620.50 g), pod length
(11.30 – 20.10 cm), pod width (6.37 – 8.90 cm), pod thickness (0.73 – 1.65 cm), number of beans
per pod (20 - 52), weight of beans per pod (27.33–119.67 g) and dry weight of hundred beans

Corresponding author: Email: sobolekky@hotmail.com

(52.33 – 115 g). Positive and significant (P < 0.001) correlation existed between pod weight and length, pod width, pod thickness and weight of beans per pod. The range of broad sense heritability was between 56.13 (number of beans per pod) to 81.76 (dry weight of hundred beans). About 86% of the total variation was explained by the first three principal component axes and four distinct groups emerged from the clustering technique. Results show significant (P<0.05) intra-cluster variability of the seven traits and that choosing genotypes G3 (T65/7 x T9/15), G5 (P7 x T60/887), G6 (P7 x PA150), G15 (T86/2 x T22/28) and G16 (T82/27 x T12/11) as parents in future yield improvement programmes will enhance cocoa productivity in Nigeria.

Keywords: Cocoa yield; parental genotypes; broad sense heritability; eigenvector; eigenvalue; principal component analysis.

1. INTRODUCTION

Agriculture has continued to play an important role in the provision of food, raw material for industries, employment of people and foreign exchange earnings in West Africa. Rural developmental activities have been greatly enhances by this sector of the economy. Industrial tree crops, notably cocoa, coffee, oil palm, and rubber, have dominated agricultural exports. Cocoa, especially have been of particular interest for some countries in the West Africa sub region and the global chocolate industries. West African countries, including Cote d'Ivoire, Ghana, Nigeria, Cameroon, Togo and others together accounted for more than 70% of total world cocoa production in 2006 [1]. Cocoa has historically been a key cash crop of very high economic value in the Nigerian agricultural system; providing the lead in quantity of foreign exchange to the nation. It is therefore remarked as the highest non-oil economic crop in Nigeria.

Cocoa (*Theobroma cacao* L.) is an under-storey tree crop of Amazonian origin, planted almost exclusively by smallholders in the tropics. Cocoa is among the major perennial crops worldwide and has enormous economic importance for developing countries in the humid tropics [2]. The major raw material from cocoa is the dried cocoa bean [3]. Sustainability of life has been dependent on this product, especially in some remote cocoa growing communities [4].

The utmost challenge to crop production is the maintenance of sustainable and productive yield; hence, improvement of yield is one of the major objectives of plant breeding. In Brazil, according to Dias et al. [5], the use of superior cocoa hybrids has contributed significantly to enhance cocoa productivity. For long, Nigeria cocoa production target has been increased productivity per hectare [3]. Identification of high quality and superior hybrids from the germplasm for further

breeding programme could lead to a boost in tonnage from cocoa hectares of the country.

The pod and bean characteristics of cocoa have very significant descriptive and discriminatory properties that guide selection of good cocoa genetic materials for breeding advancement [6]. Information on the potentials of cocoa genotypes is critical to selection. This study carried out field evaluation of 20 cocoa genotypes that were developed as hybrids. As a follow-up to the recent work by Adewale et al. [6], it would be necessary to have basic information on the pod and bean characteristics of the twenty genotypes. The information will be used to infer an opinion about the comparative advantages of each genotype and hence about the interrelationship among the studied traits. From the information, the distinctive potential of each genotype will also be determined for a guided selection programme.

2. MATERIALS AND METHODS

The experiment was carried out in an existing hybrid trial plot of Cocoa Research Institute of Nigeria (CRIN) Ibadan, Nigeria. The experimental design in the plot was Randomized Completely Block Design with six replications. The plot size was ten trees per genotypes. Planting was done at a spacing of 3 m by 3 m. Table 1 has the list of the 20 cocoa genotypes considered in the study. Data were collected from fifteen uniformly matured and ripe pods from each genotype, following the procedure of Adewale et al. [6]. Data was collected on the metric distance of the proximal to the distal end of the pod i.e. pod length (PdLT), pods were weighed and the weights were recorded as pod weight (PdWT), the circumferential length around the widest width of the pod were recorded as the pod width (PdWidth), pod thickness (PdThick) was determined as the difference between the inner and the outer diameter of the pod.

Moreover, the number of beans per pod (NoBP) was counted and the weight of all beans from a pod was recorded as weight of bean per pod (WtBP). The weight of one hundred dried beans was recorded as 100 beans (DW100B). All the data were subjected to the statistical analysis using the Statistical Analysis Software (SAS), version 9.2 [7]. The analysis of variance (ANOVA) was calculated using the PROC GLM procedure in SAS and means were separated using the Duncan multiple range test. Broad sense heritability was estimated as the ratio of genetic variance to the phenotypic variance from the component of the ANOVA and expressed in percentage following Toker [8], as

Broad Sense Heritability

$$(H) = \sigma^2 g / \sigma^2 p \times 100$$
$$\sigma^2 p = \sigma^2 g + \sigma^2 e$$

Where

$\sigma^2 g$ = genotypic variance
$\sigma^2 p$ = phenotypic variance
$\sigma^2 e$ = environmental variance

Heritability percentages were categorized as low, moderate and high as indicated by Elrod and Stanfield [9],

0 - 10% - Low
20 - 50% - Moderate
≥50% - High

Pearson Correlation among the seven studied traits was done using the Analyst option in SAS. Data matrix containing the means of each genotype with each respective trait were subjected to principal component analysis (PCA) and a three dimensional figure showing the spatial display of the twenty genotypes was generated.

3. RESULTS

The twenty genotypes differed significantly (P < 0.01) from each other for five (PdWT, PdLT, PdWidth, PdThick, and DW100B) out of the seven studied traits (Table 2). The mean of the twenty genotypes for the traits are: PdWT (384.18 g), PdLT (15.20 cm), PdWidth (7.75 cm), PdThick (1.20 cm) and DW100B (83.63 g). The coefficients of variation for the seven traits ranged between 8.15 (PdWidth) and 33.48 (WtBP).
From Table 3, G10 gave the highest mean value for pod weight (620.50 g), pod length (20.10 cm),

pod width (8.90 cm), pod thickness (1.65 cm) and number of bean per pod (46.67). The highest weight of bean per pod (119.67 g) was by G15. G22 which gave the highest value for 100 dry bean weight (115 g), however, had the least mean value for pod weight, pod width, number of beans per pod and weight of bean per pod. G11, G2, and G24 gave the least mean for pod length, pod thickness and dry weight of hundred beans respectively (Table 3).

Table 1. Description of the twenty genotypes used in the study

S/N	Code	Pedigree
1	G2	T12/11 x N38
2	G3	T65/7 x T9/15
3	G4	PA150 x T60/887
4	G5	P7 x T60/887
5	G6	P7 x PA150
6	G7	T65/7 x T57/22
7	G8	T53/5 x N38
8	G9	T65/7 x N38
9	G10	T53/5 x T12/11
10	G11	T65/35 x T30/13
11	G12	T86/2 x T9/15
12	G13	T9/15 x T57/22
13	G15	T86/2 x T22/28
14	G16	T82/27 x T12/11
15	G17	T86/2 x T16/17
16	G18	T65/7 x T53/8
17	G19	T65/7 x T101/15
18	G22	T101/15 x N38
19	G23	T82/27 x T16/17
20	G24	T86/2 x T57/22

In this study, broad sense heritability ranged from 56.13 (NoBP) to 81.76 (DW100B). Most of the traits, had very high (>70%) heritability (Table 4).

The relationship among the seven traits was presented in Table 5. The pod weight had positive and significant (P<0.001) correlation with pod length (r = 0.732), pod width (r = 0.875), pod thickness (r = 0.677) and WtBP (r = 0.733). On the other hand, significant (P≤0.05) and positive correlation existed between PdLT and PdWidth, PdThick and WtBP at r = 0.704, 0.673 and 0.476 respectively (Table 5). Other significant correlation was observed between width of pod and its thickness (r = 0.668), podwidth and WtBP (r = 0.616) and the number of beans with WtBP (r = 0.694).

From Table 6, the total genetic variation among the 20 cocoa genotypes was accounted for by seven Principal Component (PC) axes, with variance proportions ranging from 55.62% (PC1) to 0.84% (PC6). The eigenvalues for each axes followed the descending trend as the variance proportion. The total variation (85.61%) among the twenty genotypes as explained by the first three PC axes were 55.62%, 15.8% and 14.18% respectively. By magnitude prominent traits with high eigenvector loadings in PC1 were: pod weight (0.47), length (0.43) and width (0.45) were most prominent in PC1. Number of beans per pod (0.65) and weight of beans per pod (0.51) were most discriminatory in PC2 while dry weight of hundred beans had the highest eigenvector (0.98) in PC3.

The display of the twenty genotypes in the tri-dimensional plane was presented in Fig. 1. Four distinct clusters were identifiable. The population of genotypes in each cluster was: I (5), II (2), III (3) and IV (4). G2, G10, G12, G17, G22 and G24 were loosely separated in the plane as nearer neighbours to the four simple clusters. G10 had highest mean performance for five (pod weight, pod length, pod width, pod thickness and number of beans per pod) out of the seven studied traits. With respect to the five traits, G12 and G24 had quite high and closer performance to G10. The least mean for pod weight, pod width, number of beans per pod and weight of beans per pod occurred in G22 while the same had the highest mean for an hundred dry bean weight. G2 had the least mean for pod thickness.

Table 7 presents the intra cluster information of each cluster with respect to the seven traits. The genotypes in cluster I had the highest mean for pod weight (479.83 g), length (16.61 cm), width (8.16 cm) and thickness (1.13 cm). The genotypes in cluster III had higher mean number of beans per pod (48), weight of beans per pod (110 g) and 100 bean weight (94 g). The least pod thickness (0.87 cm) also occurred in the cluster (Table 7). In this study, cluster IV was prominent for: lowest PdWT (321.42 g), PdLT

(14.54 cm), PdWidth (7.53 cm), NoBP (36), WtBP (67.25 g) and DW100B (75.34 g).

4. DISCUSSION

The cocoa genotypes considered in this study exhibited significant variation for the studied pod characteristics. This proffers possibilities for subsequent breeding programme and opportunities. Vodouhe et al. [10] remarked that plant genetic resources are the most valuable resource to the world. Available variations within genetic resources offer the needed materials to tackle emerging agro-ecological problems, forms basis for parental selection and guarantees subsequent breeding success. Breeding progress is dependent on the proportion of the total variation that is genetic. The high heritability observed for the seven traits in this study indicates that there was high influence of the genetic component of variance on the phenotypic expression of the traits by the twenty genotypes. Genetic parameters and character associations provide information about expected response of various characters; this information is important for selection and development of breeding procedures [11]. Knowledge of the simple, phenotypic and genotypic correlations between important characters with breeding values is necessary for planning; it will facilitate simultaneous breeding for inheritance and aids understanding of trait association [12,13]. Simultaneous trait selection could be facilitated by the observed positive relationship of the pod weight with its length, width, thickness and the weight of bean per pod, whereby the selection of genotypes with heavier pods equally means a selection for genotypes with higher bean weight. An initial study [6] observed strong and significant positive correlation among some of the traits in the present study. In this study, pod length, width, thickness and the weight of the wet beans are indices for pod weight determination. G10 positively and remarkably demonstrated this

Table 2. Analysis of variance for the yield characters

Source of variation	DF	PdWT	PdLT	PdWidth	PdThick	NoBP	WtBP	DW100B
Genotype	19	38419.85***	15.21**	1.30***	0.12***	153.89	1274.85	687.05***
Error	38	8416.32	5.37	0.40	0.03	93.00	778.94	78.70
Mean		384.18	15.20	7.75	1.06	38.42	83.36	83.63
CV%		23.88	15.25	8.15	15.38	25.10	33.48	10.61

PdLT-Pod length, PdWT-Pod weight, PdWidth- Pod width, PdThick-Pod thickness, NoBP- Number of beans per pod, WtBP- Weight of bean per pod, and DW100B- Dry weight of hundred beans

Table 3. Mean performance of twenty genotypes for seven characters

Genotypes	PdWT	PdLT	PdWidth	PdThick	DW100B
G2	306.83^{fgh}	14.13^{bcde}	7.03^{def}	0.73^{f}	62.00^{gh}
G3	407.00^{bcdef}	17.40^{abc}	8.17^{abcd}	1.20^{bc}	94.67^{bc}
G4	341.83^{defgh}	15.90^{abcd}	7.60^{bcde}	1.03^{bcdef}	74.00^{efg}
G5	506.50^{abcd}	15.33^{bcde}	8.30^{abc}	1.00^{bcdef}	100.67^{ab}
G6	437.33^{bcdef}	17.20^{abc}	8.33^{abc}	1.07^{bcde}	93.33^{bcd}
G7	317.67^{efgh}	15.37^{bcde}	7.17^{cdef}	1.03^{bcdef}	72.67^{fg}
G8	294.50^{fgh}	13.43^{cde}	7.40^{cdef}	1.20^{bc}	78.00^{cdefg}
G9	331.67^{defgh}	13.47^{cde}	7.93^{abcd}	1.13^{bcd}	76.67^{defg}
G10	620.50^{a}	20.10^{a}	8.90^{a}	1.65^{a}	74.00^{efg}
G11	352.17^{cdefg}	11.30^{e}	7.90^{abcd}	1.03^{bcdef}	89.33^{bcdef}
G12	394.67^{bcdef}	17.60^{abc}	7.73^{abcd}	1.23^{b}	75.33^{efg}
G13	363.83^{cdefg}	14.07^{bcde}	7.70^{abcd}	0.83^{def}	94.67^{bc}
G15	555.67^{ab}	14.70^{bcde}	7.80^{abcd}	1.20^{bc}	82.67^{cdef}
G16	492.67^{abcde}	18.43^{ab}	8.20^{abcd}	1.17^{bc}	100.00^{ab}
G17	213.67^{gh}	12.23^{de}	6.47^{ef}	0.80^{ef}	64.67^{gh}
G18	371.83^{cdefg}	15.03^{bcde}	7.90^{abcd}	0.97^{bcdef}	88.00^{bcdef}
G19	289.17^{fgh}	14.70^{bcde}	7.57^{cde}	0.97^{bcdef}	90.67^{bcde}
G22	175.40^{h}	12.00^{de}	6.37^{f}	0.93^{bcdef}	115.00^{a}
G23	382.83^{bcdefg}	14.70^{bcde}	7.60^{bcde}	0.90^{cdef}	94.00^{bc}
G24	529.83^{abc}	16.97^{abc}	8.83^{ab}	1.20^{bc}	52.33^{h}

Means followed by the same letter(s) are not significantly different according to DMRT (P<0.05)
NB: PdLT-Pod length, PdWT-Pod weight, PdWidth- Pod width, PdThick-Pod thickness, NoBP- Number of beans per pod, WtBP- Weight of bean per pod, and DW100B- Dry weight of hundred beans

Table 4. Heritability for the seven characters in *T. cacao*

Characters	Heritability
PdWT	72.17
PdLT	71.03
PdWidth	73.00
PdThick	75.00
NoBP	56.13
WtBP	56.89
DW100B	81.76

NB: PdLT-Pod length, PdWT-Pod weight, PdWidth- Pod width, PdThick-Pod thickness, NoBP- Number of beans per pod, WtBP- Weight of bean per pod, and DW100B- Dry weight of hundred beans

for pod weight, length, thickness and the number of beans per pod. Adewale et al. [14] recently stated that heavier cocoa pods in most cases contain many beans and or few bigger beans. Moreover, the weight of the wet beans is a function of the number of beans, the length and width of the pods. The mode of contribution of the component traits to pod weight and wet bean weight deserves investigation through path coefficient analysis, to ascertain direct or indirect contribution.

The dry bean is the most economic part of cocoa [3]. Selection focused on genotypes with higher wet and dry bean weight will enhance greater production of cocoa. From this study, the significantly higher 100 dry bean weight value for G22 (T101/15 x N38); despite the fairly low pod weight, pod width, number and weight of beans per pod, must have been due to higher individual bean weight arising from low moisture but high dry matter content. One of the major goals of cocoa breeding is to increase the number of beans per pod and the bean weight. G10 (T53/5 x T12/11) and G22 (T101/15 x N38) positively offer these two desirable traits. A hybridization programme between the two may produce some outstanding results. An outstanding significant quality in G2 (T12/11 x N38) is the low pod thickness. Except for the importance of pod husk utilization for organic manure composting [15] for soil fertility enhancement; thinner rind of the pod husk is a desirable trait in cocoa breeding programme. Selection of genotypes with such quality in a hybridization programme may lead to physiological and genetic advances in appropriate assimilate distribution of the sink to the most economic component of the pod (i.e. the beans). Moreover, intra-cluster selection of parents and hybridization within cluster III could lead to the generation of heterotic hybrids with high quality (number and weight) bean traits.

Table 5. Pearson correlation coefficient of the seven yield trait

	PdWT	PdLT	PdWidth	PdThick	NoBP	WtBP
PdLT	0.732***					
PdWidth	0.875***	0.704***				
PdThick	0.677***	0.673***	0.668***			
NoBP	0.389^{ns}	0.389^{ns}	0.328^{ns}	0.236^{ns}		
WtBP	0.733***	0.476*	0.616**	0.234^{ns}	0.694***	
DW100B	-0.099^{ns}	-0.116^{ns}	-0.078^{ns}	-0.129^{ns}	-0.126^{ns}	-0.065^{ns}

NB: PdLT-Pod length, PdWT-Pod weight, PdWidth- Pod width, PdThick-Pod thickness, NoBP- Number of beans per pod, WtBP- Weight of bean per pod, and DW100B- Dry weight of hundred beans

Table 6. Eigenvalues, variance proportions and eigenvectors showing the prominence of each trait to each PC axes

PC-Axes	Eigenvalue	Variance proportion	Cumulative	Eigenvectors of seven yield traits						
				PdWT	PdLT	Pdwidth	PdThick	NoBP	WtBP	DW100B
PC1	3.89	55.62	55.62	0.47	0.43	0.45	0.37	0.30	0.39	-0.08
PC2	1.11	15.80	71.43	-0.93	-0.21	-0.19	-0.48	0.65	0.51	-0.04
PC3	0.99	14.18	85.61	0.08	-0.01	0.09	-0.07	-0.05	0.12	0.98
PC4	0.50	7.14	92.74	-0.34	0.30	-0.34	0.45	0.57	-0.36	0.16
PC5	0.30	4.27	97.01	0.09	-0.83	0.14	0.52	0.16	0.01	0.02
PC6	0.15	2.16	99.16	0.37	0.03	-0.78	0.26	-0.25	0.36	0.01
PC7	0.06	0.84	100.00	-0.71	0.08	0.11	0.30	-0.27	0.56	-0.01

Table 7. Cluster means showing intra-cluster variability of the seven traits

	PdWT	PdLT	PdWidth	PdThick	NoBP	WtBP	DW100B
Cluster I	479.83	16.61	8.16	1.13	38.53	98.10	94.27
Cluster II	337.72	13.68	7.79	0.99	40.33	78.00	89.33
Cluster III	373.33	14.39	7.65	0.87	47.67	110.08	94.34
Cluster IV	321.42	14.54	7.53	1.10	35.83	67.25	75.34

NB: PdLT-Pod length, PdWT-Pod weight, PdWidth- Pod width, PdThick-Pod thickness, NoBP- Number of beans per pod, WtBP- Weight of bean per pod, and DW100B- Dry weight of hundred beans

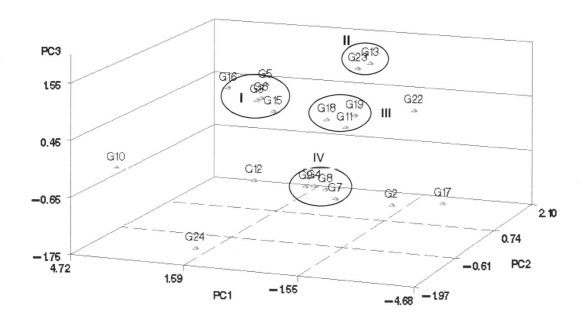

Fig. 1. Three dimensional representation revealing similarities among the twenty genotypes of cocoa

Furthermore, genotypes with comparative advantage for the economic trait (100 dry bean weight) are: G5 (P7 x T60/887), G13 (T9/15 x T57/22), G16 (T82/27 x T12/11), G22 (T101/15 x N38) and G23 (T82/27 x T16/17). Owing to the long protocol of varietal development, selection of these genotypes for further evaluation over years and locations as sexual or clonal planting material may eventually lead to their release to farmers in a short term breeding programme. In conclusion, the use of the above identified genotypes as parents in future breeding programmes will enhance the productivity of cocoa in Nigeria.

5. CONCLUSION

The findings of this study identified five cocoa genotypes (G5, G13, G16, G22 and G23) with significant dry bean quality. Their selection as parents for subsequent hybrid development may result in the generation of progenies with higher and significant heterotic quality for dry bean weight.

COMPETING INTERESTS

Authors have declared that no competing interests exist.

REFERENCES

1. ICCO. Annual Report. 2007;43
2. Ryan D, Bright GA, Somarriba E. Damage and yield change in cocoa crops due to harvesting of timber shade trees in Talamanca, CostaRica. Agroforest Syst. 2009;77:97-106
3. Oyedokun AV, Omoloye AA, Adewale BD, Adeigbe OO, Adenuga OO, Aikpokpodion PO. Phenotypic variability and diversity analysis of bean traits of some Cocoa hybrids in Nigeria. Asian J Agric Sci. 2011;3:127-131.
4. Krauss U, Ten-Hoopen M, Hidalgo E, Martínez A, Arroyo C, García J, etc. Manejo Integrado de la moniliasis (*Moniliophthora roreri*) del cacao (*Theobroma cacao* L.) en Talamanca, Costa Rica. Agroforestería en las Américas. 2003;10:52-58.
5. Dias LAS, Marita J, Cruz CD, Barros EG, Salomao TMF. Genetic distances and its association with heterosis in Cocoa. Braz Arch Bio Tech. 2003;46:339-347.
6. Adewale BD, Adeigbe OO, Adenuga OO, Adepoju AF, Muyiwa AA, Aikpokpodion PO. Descriptive and discriminatory significance of pod phenotypic traits for diversity analysis of cocoa genotypes. J Plant Breed Genet. 2013;1:131-137.

7. SAS Institute. The SAS system for Windows. Release 9.2.SAS.Inst., Gary, NC, USA; 2007.

8. Toker G. Estimate of broad sense heritability for yield and yield criteria in faba bean (*Vicia faba* L.). Hereditas. 2004;140; 222-225

9. Elrod S, Stanfield W. Genetics. 4th ed. Tata McGraw Hill, New Delhi; 2002.

10. Vodouhe SR, Atta-Krah K, Achigan-Dako GE, Eyog-Matig O, Avohou H. Plant genetic resources and food security in West and Central Africa. Regional conference. 26–30 April 2004. Bioversity International, Rome, Italy. 2007;11.

11. Gopal J. Genetic parameters and character association for family selection in potato breeding programmes. J Genet Breed. 2001;55:201–208.

12. Omokhafe KO, Alika JE. Clonal variation and correlation of seed characters in *Hevea brsiliensis* Muell. Agr Ind Crops Products. 2004;19:175–184.

13. Kaushik N, Kumar K, Kumar S, Kaushik N, Roy S. Genetic variability and divergence studies in seed traits and oil content of Jatropha (*Jatropha curcas*) accessions. Biom Bioener. 2007;31:497–502.

14. Adewale BD, Adeigbe OO, Sobowale IO, Dada SO. Breeding value of cocoa (*Theobroma cacao* L.) for pod and bean traits: A consequential advance in Nigerian cocoa breeding program. Not Sci Biol. 2014;6(2):214-219.

15. Agbeniyi SO, Oluyole KA, Ogunlade MO. Impact of cocoa pod husk fertilizer on cocoa production in Nigeria. World J. Agric. Sci. 2011;7:113–116.

13

Morphological and SSR Assessment of Putative Drought Tolerant M₃ and F₃ Families of Wheat (*Triticum aestivum* L.)

A. M. M. Al-Naggar[1*], Kh. F. Al-Azab[2], S. E. S. Sobieh[2] and M. M. M. Atta[1]

[1]Department of Agronomy, Faculty of Agriculture, Cairo University, Giza, Egypt.
[2]Department of Plant Research, Nuclear Research Center, Atomic Energy Authority, Inshas, Egypt.

Authors' contributions

This work was carried out in collaboration between all authors. Author AMMAN designed the study, performed the statistical analysis, wrote the protocol, and wrote the first draft of the manuscript. Authors KFAA and SESS managed the experiments and analyses of the study. Author MMMA managed the literature searches. All authors read and approved the final manuscript.

Editor(s):
(1) Mahalingam Govindaraj, ICRISAT, Patancheru, India.
(2) Giuliana Napolitano, Department of Biology (DB), University of Naples Federico II, Naples, Italy.
Reviewers:
(1) Klára Kosová, Department of Plant Genetics, Breeding and Crop Quality, Crop Research Institute, Prague, Czech Republic.
(2) Shilpi Srivastava, Department of Botany, D.D.U. Gorakhpur University, Gorakhpur, India.

ABSTRACT

In an attempt to develop drought tolerant genotypes of bread wheat, two procedures, i.e., mutation and hybridization were used to induce new genetic variation. Selection for high grain yield/plant (GYPP) and other desirable traits was practiced in the M₂ populations of 7 gamma irradiated genotypes and F₂ populations of 15 diallel crosses among 6 genotypes of wheat under well watering (WW) and water stress (WS) conditions. Progenies of these selections (53 M₃ and 109 F₃ families) and their seven parents were evaluated in the field under WW and WS. Significant yield superiority of twelve families (7 M₃'s and 5 F₂'s) over their original and better parents, respectively under WS reached 74.71% (SF9). These putative drought tolerant families were assessed on the DNA level using SSR analysis. Fifteen SSR primers were used for PCR amplification of the genomic DNA of these 12 selections and their parents. The SSR analysis proved that the 12 families are genetically different from their 7 parents, with an average polymorphism of 86.67%. The genetic similarities (Gs) ranged from 30% to 88%. Both mutants SF3 and SF4 exhibited very low Gs (42 and 40%, respectively) with their common parent (Giza-168), indicating that gamma rays were very effective in changing the genetic background of Giza-168 towards high GYPP under WS conditions. SSR assay permitted the identification of seven unique bands (5 positive and 2 negative) for three drought tolerant wheat genotypes (SF3, SF4 and Aseel-5). These bands might be considered useful as markers associated with drought tolerance in bread wheat breeding programs.

Keywords: Putative mutants; transgressive segregants; bread wheat; drought tolerance; SSR markers; genetic similarity.

1. INTRODUCTION

At present, the average annual consumption of wheat grains in Egypt is about 14 million tons, while the average annual local production is about 8 million tons with an average grain yield of 18.0 ardab/feddan (6.43 t/ha) [1]. The gap between annual local production and consumption is about 6 million tons, which is imported from Russia, France and other countries. To achieve self sufficiency, the area cultivated with wheat should be increased, which is possible only in the North coast and Egyptian deserts. But the soil in these areas is sandy with low water holding capacity and thus exposes wheat plants to drought stress. Such drought stress causes great losses in wheat yield and its components [2,3]. Using drought tolerant wheat cultivars that consume less water, and can tolerate soil water deficit could solve this problem.

Wheat breeders always search for broad genetic variation to start a successful breeding program for improving the trait of interest. Unfortunately, with present distribution of improved high yielding pure line cultivars in all of the world's wheat growing areas, selection from established cultivars would rarely isolate a new genotype [4]. Gamma rays proved to be effective in broadening genetic variability of wheat cultivars for grain yield and its components, helping plant breeders to practice an efficient selection in the M_2 and next mutated generations [5-8]. In a little less than a century, mutation breeding programs resulted in developing more than 3200 crop varieties that are being grown all over the world; of which 254 mutant wheat varieties were developed by physical mutagens and mutants induced via gamma rays have been obtained in bread wheat for resistance to drought leading to the release of 26 varieties worldwide [9].

Hybridization procedure between diverse genotypes is used to create hybrid populations with wide genetic variation, from which new recombinations of genes may be selected [10]. Selection from segregating generations of wheat hybrid combinations succeeded to develop new genotypes that possess adaptive traits of drought tolerance, such as early maturity [12], glaucouness [7,12] and high grain yield/plant under water deficit conditions [13,14].

Molecular markers have been proven to be more powerful tools in the assessment of genetic variation and elucidation of genetic relationships within and among species than the morphological and biochemical markers, which may be affected by environmental factors and growth practices [15,16]. A wide variety of DNA-based markers has been developed in the past few years. Simple sequence repeats (SSRs) are present in the genome of all eukaryotes and consist of several repeats to over hundreds of nucleotide motif and flanked by sequence that can be used as primers, so they are more specific than RAPDs [17]. SSRs offer a potentially attractive combination of features that are useful as molecular markers. First, SSRs have been reported to be highly–polymorphic and highly informative in plants, providing many different closely related individuals [18]. Second, SSRs can be analyzed by a rapid, technically simple, and inexpensive PCR-based away that requires only small quantities of DNA. Third, SSRs are co-dominantly inherited and reveal simple Mendelian segregation has been observed. Finally, SSRs are both abundant and uniformly dispersed in plant genomes [18,19]. Many investigators concluded that SSR molecular markers are significantly associated with wheat traits related to salinity tolerance [20] and drought tolerance [21-27].

The present investigation was carried out in an attempt to develop new wheat genotypes (mutants via gamma rays selected from M_2 populations and transgressive segregants selected from F_2 populations of hybrid combinations) tolerant to water stress conditions.

1.1 The Objectives were to

(i) evaluate the putative mutants (M_3) and transgressive segregants (F_3) families along with their parents for drought tolerance in the field, (ii) assess the genetic diversity between the best selections (seven M_3 mutants and five F_3 segregants) and their parents on the DNA level using SSR analysis and (iii) identify unique molecular markers for drought tolerant selections and/ or parents.

2. MATERIALS AND METHODS

This investigation was carried out during four successive wheat growing seasons (2008/2009 through 2011/2012) at the Experimental Farm and Molecular Genetics Laboratory of the Plant Research Department, Nuclear Research Center, Inshas, El-Sharkyia Governorate. The latitude and longitude of the experimental farm are 30º 24` N and 31º 35` E, respectively, while the altitude is 20 m above the sea level.

Six cultivars of bread wheat (*Triticum aestivum* L.), *i.e.*, Sids-4, Sakha-61, Aseel-5, Sakha-93, Giza-168 and Sahel-1 and the experimental line Maryout-5 were used in the present study. Name, pedigree, origin and important traits of these genotypes are presented in Table (1). The six genotypes, *viz.* Sids-4 (P1), Sakha-61 (P2), the line Maryout-5 (P3), Aseel-5 (P4), Sakha-93 (P5) and Giza-168 (P6) were grown in 2008/2009 season. All possible diallel crosses (excluding reciprocals) were made among the six parents, and seeds of 15 direct F_1 crosses were obtained. F_1 seeds from each of the 15 crosses were sown in the field on 20 Nov. 2009 under well watering conditions in separate plots. Each plot consisted of 6 rows, 3 m long and 30 cm wide; with hills spaced 10 cm apart (plot size = 1.8 m^2). At maturity F_2 seeds of each cross were separately harvested and kept for use in the third season (2011/2012).

Seeds of each of the seven parents irradiated with a selected dose of gamma rays (350 GY) determined by a preliminary experiment, were immediately sown on 20 Nov., 2009 in separate plots to obtain M_1 plants of each bulk. Each plot consisted of 30 rows; each row was 4 m long and 30 cm wide. Spaces between each two plants were 10 cm in each row. The plants were left for natural self pollination. At harvest, ten kernels were taken randomly from each M_1 plant (M_2 seed) and seeds from each bulk were blended to represent seed of the respective M_2 bulk. These seeds of M_2 bulks were kept for use in experiments of the third season (2010/2011). The recommended cultural practices for wheat production at Inshas were followed in M_1 and F_1 generations.

Seeds of the 15 F_2's and 7 M_2's were sown on 25 Nov., 2010 in the field under water stress (WS) and well watering (WW) in separate plots. Each plot consisted of 18 rows, 3 m long and 30 cm wide; with hills spaced 10 cm apart (plot size =5.4 m^2). Two irrigation regimes (starting 21 days after sowing) were used, viz., irrigation every 5 days (WW) and every 15 days (WS). The calculated total quantity of irrigation water for WS was 70% of that for WW and the soil at the experimental site was sandy to loamy sandy. At harvest individual plant selection, using *ca* 1% selection intensity was practiced in the same season (2010/2011), in the 15 F_2's and 7 M_2's. Selection was performed for grain yield/ plant and some other favorable traits, such as spike length, spike weight, spikes/plant, earliness, glaucousness...*etc.*, under water stress and non-stress conditions. One hundred and sixty two individual plant selections were separately harvested (53 from M_2 and 109 from F_2 populations).

2.1 The Field Experiment

A field experiment was conducted in 2011/ 2012 season to compare the selected individual genotypes with their parents. The experimental design used was a split-plot in a balanced lattice (13 x 13) arrangement with three replications. Main plots were assigned to two irrigation regimes (WW and WS) and sub-plots were devoted to 169 genotypes (162 selections + 7 parents). Each plot consisted of 4 rows, 2.25 m long and 30 cm wide; with hills spaced 10 cm apart (plot size = 2.7 m^2). Rainfall in 2010/2011 and 2011/2012 seasons were very light and intermittent with a total precipitation of 10.3 and 13.9 mm, respectively, suggesting that rainfall during the stress period was of negligible influence on moisture content of the experimental soil.

Data was recorded in the field on: 1. days to 50% heading (DTH), 2. days to 50% anthesis (DTA), 3. days to 50% physiological maturity (DTM), 4. plant height (PH), 5. spike length (SL), 6. spikes / plant (SPP), 7. grains / spike (GPS), 8. spike weight (SW), 9. 100-grain weight (100GW) and 10.grain yield / plant (GYPP). Data on traits No. 1, 2 and 3 were measured on a per plot basis.

Table 1. Pedigree, origin and the most important traits of the studied wheat genotypes

Genotype	Designation	Pedigree	Origin	Important trait
Sids-4 cv.	Sd-4	Maya"S"Mon"S"/CMH74.A592/3/Sakha8 X2SD10002-140sd-3sd-1sd-0sd	ARC – Egypt	Earliness
Sakha-61 cv.	Sk-61	Lina/RL4220//7c/Yr"S"CM 15430-25-55-0S-0S	ARC – Egypt	Earliness
Maryout-5 Line	Mr-5	Giza 162 // Bch's /4/ PI-ICW 79Su511Mr-38Mr-1Mr-0Mr	DRC – Egypt	High yielding and Salt tolerant
Aseel-5 cv.	As-5	BIG INC 08 104	ICARDA - Syria	Drought tolerant
Sakha-93 cv.	Sk-93	Sakha 92/ TR 810328 S8871-1S-2S-1S-0S	ARC – Egypt	High yielding
Giza-168 cv.	Gz-168	Mrl / Buc // Seri CM 930468M-0Y-0M-2Y-0B	ARC – Egypt	High yielding
Sahel-1 cv.	Sah-1	NS 732 / PIMA // VEERY "S"	ARC – Egypt	Drought tolerant

ARC = Agricultural Research Center, DRC = Desert Research Center, ICARDA = International Center for agricultural Research in the Dry Areas, cv. = cultivar

Data on other traits (No. 4 through 10) were measured on 30 individual plants/plot. Recorded data were subjected to the normal analysis of variance for balanced lattice design, and the least significant differences (LSD) between means were estimated according to Snedecor and Cochran [28].

2.2 Molecular Analysis

SSR analysis was used in the present study to assess the genetic diversity among the best 12 drought tolerant selections (7 putative mutants and 5 transgressive segregants) and their 7 parents and to identify markers associated with drought tolerance.

Young green leaves were collected from ten days-old seedlings germinated from seeds of each genotype and quickly frozen in liquid nitrogen and then ground using mortar and pestle. Extraction of genomic DNA from these leaves was carried out according to Doyle and Doyle [29] and Sumar et al. [30].

The polymorphism among the 12 drought tolerant selections (7 putative mutants and 5 transgressive segregants) and their 7 parents was detected based on SSR analysis. A set of fifteen random primers (Table 2) chosen according to Bousba et al. [31] among the publicly available sets catalogued in the Grain Genes database (http://wheat.pw.usda.gov) as

WMC (Xwmc). Roider et al. [32] described the WMS (Xgwm) as specialized for *Triticum aestivum* and used for screening drought tolerance. These primers were synthesized by BioShop® Canada Inc. and used for SSR analysis.

The PCR master mix for the simple sequence repeat (SSR) primers consisted of 2 µl of 20 ng/µl genomic DNA template, 0.40 µl of10 µM a forward and reverse primer mixture, 0.18 µl (0.9 U) of Taq polymerase, 1.20 µl of 10X buffer (10 mMTris-HCl, 50 mM KCl, 1.5 53 mM MgCl$_2$, pH 8.3), 0.96 µl of a 100 µM mixture of dNTPs and 7.26 µl of water bringing the total reaction volume to 12 µl. Reaction conditions for SSR markers were as follows: 8.33 µl ddH20, 2.4 µl 10 X reaction buffer, 0.9 µl 50 mM MgCl$_2$, 1.92 µl 2.5 mM dNTPs, 1.9 µl 1pM of 19bp M-13. The PCR master mix was carried out in a volume of 20 µl and contained 200 ng of genomic DNA, 0.2 mM of dNTPs, 10 pmol of each primer, 2.0 mM of MgCl$_2$, 50 mM of KCl, 10 mM of Tris-HCl (pH 9.0 at 25ºC), 0.1% TritonX-100 and 0.5 U of Taq DNA Polymerase. The amplification products were resolved by electrophoresis in a 1.5% agarose gel containing ethidium bromide (0.5 ug/ml) in 1 X TBE buffer at 95 volts. PCR products were visualized on UV light and photographed using a Polaroid camera. Amplified products were visually examined and the presence or absence of each size class was scored as 1 or 0, respectively.

The banding patterns generated by SSR-PCR marker analysis were compared to determine the genetic relatedness of the genotypes. Bands of the same mobility were scored as identical. The genetic similarity coefficient (GS) between two genotypes was estimated according to Dice coefficient [33] as follows: Dice formula: $GS_{ij} = 2a/(2a+b+c)$, where GS_{ij} is the measure of genetic similarity between individuals i and j, a is the number of bands shared by i and j, b is the number of bands present in i and absent in j, and c is the number of bands present in j and absent in i.

The similarity matrix was used in the cluster analysis. The cluster analysis was employed to organize the observed data into meaningful structures to develop taxonomies. At the first step, when each accession represents its own cluster, the distances between these accessions are defined by the chosen distance measure (Dice coefficient). However, once several accessions have been linked together, the distance between two clusters is calculated as the average distance between all pairs of accessions in the two different clusters. This method is called unweighted pair group method using arithmetic average (UPGMA) according to Sneath and Sokal [33].

3. RESULTS AND DISCUSSION

3.1 Field Experiment

3.1.1 Analysis of variance

Analysis of variance of the split plot experiment that included two irrigation regimes in the main plots and 169 wheat genotypes in the sub-plots (53 M_3 selected families, 109 F_3 selected families and 7 parents) for studied characters is presented in Table (3).

Results indicated that mean squares due to irrigation regimes and those due to genotypes were highly significant for all studied traits, suggesting the significant effect of both irrigation regime and genotype on such traits. Mean squares due to genotypes X irrigation regimes interaction were highly significant for all studied traits, suggesting that performance of the studied genotypes in this experiment varied with water supply, confirming the results of other workers [7,8,11,34,35].

3.1.2 Morphological assessment of the best 12 selected families

The best twelve selected families (SF) included 7 M_3 families; two (SF2 from M_2 of Sakha-93 and SF3 from M_2 of Giza-168) selected under WS, and five (SF1 from M_2 of Aseel-5, SF4 and SF5 from M_2 of Giza-168, SF6 and SF7 from M_2 of Sahel-1) selected under WW and 5 F_3 families; three (SF9, SF10 and SF11) selected under WS, from the F_2 of Sd4 X Mr5, Sk61 XAs5 and Sk61 X Sk93, respectively and two (SF8 and SF12) selected under WW, from the F_2 of Sd4 X Sk61 and Mr5 X Sk93, respectively.

Means of studied traits of the best 12 families and the 7 parental genotypes under WS and WW are presented in Table (4). On average, under WS conditions the group of the best 5 F_3 families showed the highest mean grain yield (41.2g), while the group of 7 parents exhibited the lowest grain yield (26.6g). Moreover, yield reduction due to water stress in the best M_3 and best F_3 groups (12.0 and 13.3% on average, respectively) was less than that of the parents group (17.1%). This means that, in this experiment, selection practiced in both M_2 and F_2 populations was effective in producing higher yielding families under WS than the original parents and the success of the two procedures, i.e., gamma-rays mutation induction and hybridization followed by transgressive segregation, in isolating new variants of higher drought tolerance. This conclusion was previously confirmed by Sobieh [6] and Al-Naggar et al. [7,8] for the success of mutation breeding. It is worth noting that the group of best F_3 families was, on average, earlier than the group of parents for DTH (by 5.3 days), DTA (by 3.9 days) and DTM (by 1.9 days) under WS Table (3). Comparing all the 12 best families (Table 5), it is interesting to mention that the best family in grain yield/plant under water stress was SF9 (45.6 g), followed by SF11 (44.2 g) and SF3 (42.8 g) with a very low reduction due to water stress (6.9, 6.2 and 11.2%, respectively). It is worth noting that the best three families under WS resulted from selection for high yield under water stress conditions.

The earliest M_3 family for DTM was SF6 as compared with the earliest parents Sids-4, Sakha-61 and Aseel-5, under water stress. The best M_3 and F_3 families for grain yield/plant were characterized by high value of one or more of yield components.

Table 2. Description of the SSR loci used in this study

No.	Primer	Sequence	
		Forward	Reverse
1	WMS 06	5 - CGT ATC ACC TCC TAG CTA AAC TAG - 3	5 - AGC CTT ATC ATC ATG ACC CTA CCT T - 3
2	WMS 30	5 - ATC TTA GCA TAG AAG GGA GTG GG - 3	5 - TTC TGC ACC CTG GGT GAT TGC - 3
3	WMS 108	5 - ATT AAT ACC TGA GGG AGG TGC - 3	5 - GGT CTC AGG AGC AAG AAC AC - 3
4	WMS 118	5 - GAT GGT GCC ACT TGA GCA TG - 3	5 - GAT TG TCA AAT GGA ACA CCC - 3
5	WMS 149	5 - CAT TGT TTT CTG CCT CTA GCC - 3	5 - CTA GCA TCG AAC CTG AAC AAG - 3
6	WMS 169	5 - ACC ACT GCA GAG AAC ACA TAC G - 3	5 - GTG CTC TGC TCT AAG TGT GGG - 3
7	WMC 177	5 - AGGGCTCTCTTAATTCTTGCT - 3	5 - GGTCTATCGTAATCCACCTGTA - 3
8	WMC 179	5 - CATGGTGGCCATGAGTGGAGGT - 3	5 - CATGATCTTGCGTGTGCGTAGG - 3
9	WMS 198	5 - TTG AAC CGG AAG GAG TAC AG - 3	5 - TCA GTT TAT TTT GGG CAT GTG - 3
10	WMC 235	5 - ACTGTTCCTATCCGTGCACTGG - 3	5 - GAGGCAAAGTTCTGGAGGTCTG - 3
11	WMS 304	5 - AGGAAACAGAAATATCGCGG - 3	5 - AGG ACT GTG GGG AAT GAA TG - 3
12	WMC 307	5 - GTTTGAAGACCAAGCTCCTCCT - 3	5 - ACCATAACCTCTCAAGAACCCA - 3
13	WMC 322	5 - CGCCCCACTATGCTTTG - 3	5 - CCCAGTCCAGCTAGCCTCC - 3
14	WMS 375	5 - ATTGGCGACTCTAGCATATACG - 3	5 - GGGATGTGTCTGTTCCATCTTAGC - 3
15	WMC 445	5 – AGAATAGGTTCTTGGGCCAGTC - 3	5 – GAGATGATCTCCTCCATCAGCA - 3

Table 3. Analysis of variance of split plot design for 169 genotypes including 162 selected families (53 from M_2 and 109 from F_2) and 7 parents under water stress and well watering conditions (Inshas, 2011/ 2012 season)

S.V.	d.f.	Mean squares				
		Days to heading	Days to anthesis	Days to maturity	Plant height	Spike length
Replication	2	14.1	9.3	2.8	1.1	0.0183
Watering (W)	1	1063.4**	1550.3**	943.7**	9040.4**	67.9**
Error [a]	2	1.6	5.2	7.5	1.0	0.01
Genotypes (G)	168	71.1**	101.8**	72.7**	336.3**	7.5**
G x W	168	4.0**	5.0**	0.6**	23.2**	0.4**
Error [b]	672	0.3	0.2	0.2	0.5	0.01
		Spike weight	Spikes/plant	Grains /spike	100-grain weight	Grain yield /plant
Replication	2	0.0002	0.0005	92.2	0.01	0.1
Watering (W)	1	44.2**	285.3**	9263.7**	96.0**	13576.2**
Error [a]	2	0.02	0.0005	5.8	0.003	1.0
Genotypes (G)	168	0.9**	14.7**	187.5**	1.0**	220.2**
G x W	168	0.2**	1.2**	8.6**	0.2**	24.7**
Error [b]	672	0.003	0.01	0.6	0.003	0.4

** = significant at 0.01, probability level

Table 4. Mean performance of the 12 best selected families (7 best M_3 and 5 best F_3 families) and their parents for studied wheat traits under water stress (WS) conditions (2011/ 2012 season)

Genotypes	DTH (day)	DTA (day)	DTM (day)	PH	SL	SW	SPP (No)	GPS (No)	100GW	GYPP	Red.
				(cm)	(cm)	(g)			(g)	(g)	%
Best M_3											
SF1	95	111	141	95	13.7	3.6	11.7	75	4.4	42.1	10.0
SF2	102	112	141	96	14.1	3.3	13.3	68	4.7	42.0	12.1
SF3	91	102	135	89	14.1	3.7	11.9	74	4.3	42.8	11.2
SF4	94	103	137	87	13.7	4.0	10.1	71	4.4	39.9	10.1
SF5	93	102	137	84	13.5	3.5	11.3	65	4.6	39.3	11.9
SF6	95	105	129	101	13.4	3.7	10.9	68	4.8	40.2	13.0
SF7	98	111	139	80	13.1	3.3	11.9	64	4.8	38.2	15.3
Av. (M_3)	95.4	106.6	137.0	90.3	13.7	3.6	11.6	69.3	4.6	40.6	12.0
Best F_3											
SF8	89	98	131	103	13.5	3.6	10.9	67	5.0	38.2	11.6
SF9	82	92	131	97	14.3	4.1	11.2	71	5.0	45.6	6.9
SF10	92	100	132	90	12.0	4.0	9.7	72	5.5	38.5	29.0
SF11	88	96	133	85	13.9	3.9	11.4	64	5.6	44.2	6.2
SF12	87	99	131	85	16.3	5.0	8.0	64	5.6	39.4	12.6
Av. (F_3)	87.6	97	131.6	92	14	4.1	10.2	67.6	5.3	41.2	13.3
Parents											
Sids-4	87	95	132	96	16.2	4.3	5.3	84.0	5.0	23.1	24.6
Sakha-61	92	100	132	79	10.3	3.1	8.1	63.0	4.4	24.8	17.7
Maryout-5	95	103	138	94	14.2	3.8	6.9	76.0	4.9	26.1	13.4
Aseel-5	96	101	132	92	13.1	3.4	9.1	69.0	4.6	33.3	10.6
Sakha-93	94	101	132	81	12.2	3.2	8.7	66.0	4.4	28.2	17.0
Giza-168	95	102	136	86	12.6	3.6	7.3	65.0	4.2	26.0	15.5
Sahel-1	94	107	133	100	13.3	3.3	7.5	68.0	4.8	24.7	20.8
Av. (P)	92.9	100.9	133.5	89.9	13.1	3.5	7.6	70.1	4.6	26.6	17.1
LSD $_{0.05}$	0.67	0.58	0.56	1.08	0.13	0.08	0.13	0.90	0.07	0.80	

Red. (Reduction %) = 100(GYPP under WW - GYPP under WS)/ GYPP under WW, P = Parents, Av. = Average
F_3 = best F_3 families, M_3 = best M_3 families

Practicing selection in the F_2 generation of the studied crosses resulted in a significant superiority (selection gain) over the better parent of the corresponding cross in grain yield/plant ranging from 15.48% for SF10 to 74.71% for SF9 under water stress and from 32.76% for SF12 to 60.24% for SF9 under non-stress conditions (Table 5). The SF9 selected F_3 family showed the highest selection gain under both water stress and non-stress conditions.

The five selected F_3 families (SF8, SF9, SF10, SF11 and SF12) showed significant superiority in grain yield over their better parents under both stress and non-stress conditions. These superior families in grain yield are the result of transgressive segregation and may be considered promising lines having tolerance to drought conditions. Observations on transgressive segregation in segregating hybrid generations were previously explained by several research workers, e.g., Al-Bakry et al. [36]. The results from classical genetic studies have provided fairly convincing evidence for the hypotheses that transgressive segregation can result from the complementary gene action [37].

Practicing selection for high grain yield in the M_2 populations derived from gamma radiation treatment of parent cultivars of wheat resulted in an actual progress over the corresponding original parent in GYPP ranging from 26.27 to 64.36% under WS for SF1 and SF3, respectively (Table 5). The SF3 selected M_3 family showed the highest selection gain followed by SF6 (62.62% under WS). These two M_3 families showed also superiority in SPP and in DTM, i.e., earliness of maturity.

Superiority in grain yield of the 12 best families over the Egyptian cultivar Sids-4 reached 97.8% for SF9, 91.8% for SF11and 85.7% for SF3 under water stress. The twelve selected families should further be selfed for more generation to reach complete homozygosity to be tested for their stability under a variety of water stress conditions.

3.1.3 The most important traits of the best 12 selections

SFI: It is a high yielding mutant under WS (2nd highest best M_3s) with low reduction (10.0%) due to water stress, i.e., drought tolerant. It recorded the highest number of grains/spike amongst the 7 best M_3 families (Fig. 1).

SF2: It is a high yielding mutant under WS conditions; with low yield reduction due to water stress (drought tolerant). It recorded the highest number of spikes (13.3) under water stress (Fig. 2).

Table 5. Actual progress (%) of the best selections over the original parent (from M_2's) and over the better parent (from F_2s) for DTM, SPP and GYPP under water stress (WS) and well watering (WW) conditions (2011/ 2012 season)

Best families	Pedigree	DTM		SPP		GYPP	
		WW	WS	WW	WS	WW	WS
Best M_3 families		**Progress (%) over the original parent**					
SF1	As-5-WW-PM5	6.77	7.22	21.78	28.57	25.44	26.27
SF2	Sk-93-WS-PM2	5.97	7.22	40.21	52.87	40.71	49.04
SF3	Gz-168-WS-PM2	-1.09	-0.74	53.01	63.01	56.34	64.36
SF4	Gz-168-WW-PM5	1.09	0.74	31.33	38.36	44.02	53.23
SF5	Gz-168-WW-PM6	1.09	0.74	50.60	54.79	44.66	50.92
SF6	Sh-1-WW-PM6	-2.60	-3.01	40.24	45.33	48.03	62.62
SF7	Sh-1-WW-PM7	4.83	4.51	50.00	58.67	44.50	54.53
Best F_3 families		**Progress (%) over better parent**					
SF8	Sd4XSk.61-WW-PS8	-0.37	-0.76	26.37	34.57	41.27	54.22
SF9	Sd4XMr5-WS-PS2	-1.12	-0.76	68.06	62.32	60.24	74.71
SF10	Sk61XAs5-WS-PS3	0.37	0.00	17.82	6.59	45.27	15.48
SF11	Sk61XSk93-WS-PS2	0.37	0.76	20.62	31.03	38.65	56.85
SF12	Mr5XSk93-WW-PS8	-0.75	-0.76	-2.06	-8.05	32.76	39.82

Fig. 1. The highest number of grains/spike for SF1 and SF3 as compared with their parents As-5 and Gz-168, respectively, and the longest spike for SF3

Fig. 2. The highest number of spikes for SF2 and SF7as compared with their parents Sk-93 and Sh-1, respectively

SF3: This mutant ranked first in grain yield/plant amongst the 7 best M_3 families under both WS and WW conditions; with low yield reduction due to water stress, *i.e.*, a drought tolerant family. It recorded the second largest number of grains/spike under WS and the longest spike (Fig. 1) and the earliest in DTH and DTM under WW and WS.

SF4: It is a high yielding mutant under both WW and WS; with low yield reduction due to water stress, *i.e.,* drought tolerant. It recorded the heaviest spike and grain under both irrigation regimes.

SF5: It is a high yielding mutant under WS conditions; with low reduction in GYPP due to water stress, *i.e.,* a drought-tolerant family.

SF6: It is a high yielding mutant under WS conditions, with low reduction in GYPP due to water stress, *i.e.,* a drought-tolerant family. It ranked the earliest amongst the best 12 families and the 7 parents. It recorded the heaviest grain under both irrigation regimes.

SF7: It is a high yielding M_3 family under both WW and WS conditions; with low yield reduction due to water stress. It is also characterized by the shortest plant height, the heaviest grain and the second highest in SPP (Fig. 2) amongst the 7 best selected M_3 families.

SF8: It is a transgressive segregant in the F_3 generation. It showed high GYPP under WS; with low yield reduction due to water stress. It also recorded the tallest plant (Fig. 3) and was earlier than the earliest parent.

SF9: It is a transgressive segregant in the F_3 generation. It showed the highest GYPP under WS; with the second lowest yield reduction (6.9%) due to WS, *i.e,* the 2nd most drought tolerant F_3 family. It is the earliest F_3 for DTH and DTA (Fig. 4).

SF10: It is a transgressive segregant in the F_3 generation. It recorded significantly higher yield than the best parent (Mr-5) under drought stress conditions. This family (SF10) also recorded the heaviest grain (Fig. 5) under both irrigation regimes.

SF11: It is a transgressive segregant in the F_3 generation. It is the most drought tolerant selected family; since reduction in its yield due to water stress was the lowest (6.2%). Its yield under WS ranked the second highest and amongst the 5 best F_3 families. This selected family showed the heaviest grain (Fig. 5) under both WW and WS conditions.

SF12: It is a high yielding family under WS; with low yield reduction (12.6%) due to water stress. It is characterized by the longest and heaviest spike (Fig. 5).

3.2 SSR Assessment

3.2.1 Genetic polymorphism among the 19 wheat genotypes

Fifteen SSR primers revealed discernible amplification profiles, therefore they were employed to investigate the genetic polymorphism among the 19 wheat genotypes (Table 5).

Fig. 3. The earliest maturity and tallest plant shown by SF8 as compared with the better parent Sids-4

Fig. 4. The earliest heading shown by SF9 as compared with the better parent Sids-4

Fig. 5. The longest and heaviest spike of SF12 as compared with the better parent Maryout-5 and Sakha-93

The 15 SSR primers produced 46 amplicons, out of them 42 were polymorphic and the average percentage of polymorphism was 95.65% (Table 6). The number of amplicons per primer ranged from 1 (WMS 30, WMC 235 and WMS 304) to 10 (WMC 179) with an average of 3.07 fragments/primer across the different genotypes. However, the number of polymorphic amplicons varied from 0 (WMC 235 and WMS 304) to 10 (WMC 179) with an average number of polymorphic amplicons of 2.93 fragments/primer. Thirteen out of the 15 primers exhibited 100% polymorphism, while two primers (WMC235 and WMS 304) showed no polymorphism. The size of amplified fragments varied with the different primers, ranging from 50 to 1500 bp. In this context, Naghavi et al. [38] used RAPD and SSR analyses to estimate genetic diversity among bread wheat genotypes including nineteen Iranian cultivars and two lines (Shain and Line 518). The level of polymorphism was 88% with RAPDs compared to 100% with SSRs. Abd El-Hadi [25] investigated the genetic diversity among three durum wheat cultivars and their six selected drought tolerant lines with ISSR analysis. He reported that out of 99 amplified DNA fragments, 70 were polymorphic, representing a level of 71.42% polymorphism. Moreover, Bousba et al. [31] reported that a total of 136 fragments were obtained from the 26 SSR primers and all the bands were polymorphic across all screened genotypes. They added that polymorphism information content (PIC) values ranged from 38% to 94%, with an average of 74%. The results of the present study are in good agreement with those reported in the literature and confirm that polymorphism is a general phenomenon in wheat populations resulting after irradiation with gamma rays and hybridization followed by segregating generations, as in the case of this study.

3.2.2 Identification of unique SSR markers for drought tolerance

Unique markers are defined as bands that specifically identify an accession from the other by their presence or absence. The bands that are present in one accession but not found in the others are termed positive unique markers (PUM), in contrast with the negative unique markers (NUM), which are absent in a specific genotype. These bands could be used for genotype identification [39].

As shown in Table (7), the SSR assay permitted the identification of three out of 19 wheat genotypes by unique positive and/or negative markers. These three genotypes, namely, SF3, SF4 and Aseel-5 (all are drought tolerant) are characterized by five positive unique markers, while one of them (SF4) was characterized by two negative unique markers.

The selected drought tolerant mutant (SF3) was characterized by three unique positive markers amplified by the primers WMC 177 (100 bp) and WMC 179 (800 and 1000 bp). The selected drought tolerant mutant (SF4) was characterized by one unique positive marker amplified by the primer WMC 179 (50 bp) and two negative unique markers amplified by the primers WMC 177 (200 bp) and WMC 179 (550 bp). The drought tolerant Syrian parent (Aseel-5) was characterized by one positive unique marker amplified by the primer WMS 198 (100 bp). The remaining 16 wheat genotypes did not exhibit any unique marker. The highest number of unique markers (four) was amplified by the primer WMC 179 (3 positive and one negative) followed by the primer WMC 177 (two unique markers; one positive and one negative). The size of these unique markers ranged from 50 to 1000 bp.

In this context, Moghaieb et al. [40] determined the genotype specific SSR markers in nine bread and pasta wheat genotypes. They reported that 13 markers can be considered as a useful marker for screening for salt tolerance in these wheat genotypes. Abd El-Hadi [25] reported that in durum wheat, ISSR analysis showed four genotype–specific markers for the drought tolerant putative line S_3 that show a significant increase in grain yield/plant over their parents under drought stress conditions. Using SSR analysis, we were able to identify seven unique bands in some drought tolerant wheat genotypes. These bands might be considered useful as markers associated with drought tolerance in bread wheat breeding programs. Further experiments need to be achieved to determine the linkage between the genotype–specific SSR markers used in the present study and gene(s) for drought tolerance in the studied bread wheat genotypes. The present results support the idea that SSR analysis can provide a fast detection of species-specific markers linked to drought stress tolerance in bread wheat.

3.2.3 Genetic similarities based on SSR analysis

The scored data from the SSR analysis in this study were used to compute the similarity matrices according to Dice coefficient [33]. As shown in Table (8) the genetic similarity ranged from 30% (between SF4 and each of Sakha-61 and Maryout-5) to 88% (between SF7 and each of SF1 and SF6). High genetic similarity between SF6 and SF7 is attributed to the fact that both of them were derived from the Sahel-1 cultivar irradiated by 350 Gy gamma rays.

The results of this investigation indicated that all the twelve selected drought tolerant families differ from their parents at the DNA level where

the average of genetic similarity (GS) between selections and their parents was about 68%. The mutants SF3 and SF4 exhibited very low genetic similarity with their common parent Giza-168 (42 and 40%, respectively), indicating that gamma rays were very effective in changing the genetic background of Giza-168in a positive direction, i.e., towards high GYPP under WS conditions. In this context, Abd El-Hadi [25] reported that the genetic similarity between six selected putative durum wheat mutants (derived via gamma rays) and their parents, based on ISSR analysis, ranged from 12.7 to 87.4%. Munir et al. [20] also reported that genetic similarity coefficients for SSR markers between 18 salt tolerant wheat accessions ranged from 45 to 95%.

3.2.4 Cluster analysis as revealed by SSR

The Dice SSR-based coefficients of genetic similarity among the 19 wheat genotypes were employed to develop a dendrogram using the UPGMA method (Fig. 6). The dendrogram separated the selected F_3 family (SF4) from the other wheat genotypes, which formed a cluster in which the selected F_3 family SF3 was separated from the remaining 17 genotypes. This demonstrates the distinctiveness of the genetic background of these two genotypes (SF3 and SF4) from all the other genotypes.

The remaining 17 genotypes were divided into three main groups. The first group was divided into two sub-groups; the first sub-group separated Sakha-61 from two other genotypes (Maryout-5 and Aseel-5) and the second sub-group was divided into two classes; one of which included two genotypes (Sids-4 and SF2) and the second class separated SF6 from the other two genotypes (SF1 and SF7).

The second group separated SF8 (in one sub-group) from 4 other genotypes (in another sub-group); the latter sub-group separated SF5 from three other genotypes in a separate class; this class separated SF10 from the other two genotypes (Giza -168 and SF9) in one sub-class. The third group separated SF12 (in one sub-group) from the remaining 3 genotypes in another sub-group. The second sub-group separated SF11 in one class from the remaining two genotypes (Sakha-93 and Sahel-1) in another class.

Table 6. Number of monomorphic and polymorphic amplicons and percentage of polymorphism, as revealed by SSR primers for 19 wheat genotypes (12 selected families and their 7 parents)

Primer	Total no of amplicons	No of mono- morphic amplicons	No of poly- morphic amplicons	Polymorphism (%)
WMS 06	2	0	2	100
WMS 30	1	0	1	100
WMS 108	6	0	6	100
WMS 118	3	0	3	100
WMS 149	3	0	3	100
WMS169	2	0	2	100
WMC 177	2	0	2	100
WMC 179	10	0	10	100
WMS 198	5	0	5	100
WMC 235	1	1	0	0
WMS 304	1	1	0	0
WMC 307	2	0	2	100
WMC 322	2	0	2	100
WMS 375	2	0	2	100
WMC 445	4	0	4	100
Total	46	2	44	
Average	3.07	0.13	2.93	95.65

Table 7. Unique positive and negative SSR markers generated for 19 wheat genotypes (12 selected families and their 7 parents), marker size (bp) and total number of markers identifying each genotype

Genotype	Positive unique markers		Negative unique markers		
	Primer (band size/bp)	Total no.	Primer (band size/bp)	Total no.	Grand total
Sids-4	-		-		
Sakha-61	-		-		
Maryout-5	-		-		
Asseel-5	WMS 198 (100)	1	-		1
Sakha-93	-		-		
Giza-168	-		-		
Sahel-1	-		-		
SF1	-		-		
SF2	-		-		
SF3	WMC 177 (100), WMC 179 (800, 1000)	3	-		3
SF4	WMC 179 (50)	1	WMC 177 (200), WMC 179 (550)	2	3
SF5	-		-		
SF6	-		-		
SF7	-		-		
SF8	-		-		
SF9	-		-		
SF10	-		-		
SF11	-		-		
SF12	-		-		
Total		5		2	7

Table 8. Genetic similarity (GS) matrices among the nineteen wheat genotypes (12 selected families and 7 parents)

Genotype	Sd-4	Sk-61	Mr-5	As-5	Sk-93	Gz-168	Sah-1	SF1	SF2	SF3	SF4	SF5	SF6	SF7	SF8	SF9	SF10	SF11
Sd-4	1.00																	
Sk-61	0.68	1.00																
Mr-5	0.63	0.82	1.00															
As-5	0.69	0.80	0.84	1.00														
Sk-93	0.73	0.72	0.67	0.72	1.00													
Gz-168	0.74	0.82	0.78	0.77	0.73	1.00												
Sah-1	0.63	0.73	0.81	0.83	0.83	0.74	1.00											
SF1	0.77	0.75	0.84	0.86	0.71	0.84	0.76	1.00										
SF2	0.85	0.68	0.71	0.78	0.69	0.70	0.63	0.81	1.00									
SF3	0.57	0.50	0.51	0.45	0.36	0.42	0.43	0.48	0.51	1.00								
SF4	0.69	0.30	0.30	0.39	0.44	0.40	0.33	0.46	0.50	0.37	1.00							
SF5	0.73	0.71	0.73	0.72	0.72	0.81	0.69	0.78	0.68	0.41	0.50	1.00						
SF6	0.71	0.65	0.75	0.74	0.70	0.75	0.71	0.85	0.76	0.50	0.49	0.73	1.00					
SF7	0.80	0.74	0.75	0.81	0.78	0.75	0.71	0.88	0.84	0.45	0.54	0.82	0.88	1.00				
SF8	0.70	0.68	0.62	0.57	0.61	0.70	0.54	0.59	0.65	0.44	0.34	0.73	0.62	0.67	1.00			
SF9	0.79	0.72	0.65	0.68	0.64	0.88	0.65	0.76	0.74	0.42	0.47	0.81	0.70	0.74	0.80	1.00		
SF10	0.80	0.73	0.67	0.69	0.73	0.79	0.67	0.77	0.75	0.46	0.44	0.77	0.67	0.76	0.81	0.84	1.00	
SF11	0.70	0.77	0.70	0.73	0.80	0.78	0.81	0.75	0.61	0.39	0.42	0.76	0.71	0.75	0.70	0.77	0.78	1.00
SF12	0.60	0.67	0.65	0.64	0.71	0.64	0.72	0.67	0.64	0.43	0.31	0.71	0.69	0.69	0.73	0.71	0.68	0.79

Sd-4= Sids-4, Sk-61= Sakha-61, Mr-5= Maryout-5, As-5= Aseel-5, Sk-93= Sakha-93, Gz-168= Giza-168, Sah-1= Sahel-1, SF1 to SF12= Selected families

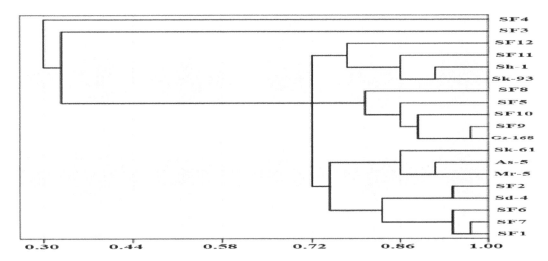

Fig. 6. Dendrogram for the nineteen wheat genotypes (12 selected families and 7 parents) constructed from SSR data using (UPGMA) according to Dice coefficients

4. CONCLUSION

The present investigation concluded that exposing some wheat cultivars and lines to gamma rays at a dose of 350 GY could induce a number (7) of putative mutants, that showed significant superiority in grain yield over the best parents reaching 64.36% for SF3 under water sress conditions. Also, some transgressive segregants (5) selected from F_2 generation of hybrids between wheat cultivars and lines showed significant superiority in grain productivity under water deficit reaching 74.71% for SF9. These new genotypes were considered drought tolerant. Molecular assessment of these mutants, transgressive segregants and their parents by SSR analysis proved the genetic dissimilarity among these new genotypes and their parents, indicating the efficiency of the two breeding methods used in this study in inducing drought tolerant genotypes.SSR assay permitted the identification of seven unique bands (5 positive and 2 negative) for three drought tolerant wheat genotypes (SF3, SF4 and Aseel-5). These bands might be considered useful as markers associated with drought tolerance in bread wheat breeding programs. Further experiments need to be achieved to determine the linkage between the genotype–specific SSR markers used in the present study and gene(s) for drought tolerance in the studied bread wheat genotypes.

ACKNOWLEDGMENTS

We would like to acknowledge the Atomic Energy Authorities in Egypt for exposing wheat seeds to gamma rays and offering the laboratory facilities for carrying out the SSR analysis. Thanks are also due to Dr. Reda Shabana and Dr. Mazhar Fawzi, Professors of Plant Breeding, Faculty of Agriculture, Cairo University for revising the manuscript of this article.

COMPETING INTERESTS

Authors have declared that no competing interests exist.

REFERENCES

1. MALR/ARE. Ministry of Agriculture and Land Reclamation, Arab Republic of Egypt. Agricultural Statistics; 2012.

2. Clarke JM, DePauw RM, Townley-Smith TF. Evaluation of methods for quantification of drought tolerance in wheat. Crop Sci. 1992;32(3):723-728.

3. Mirbahar AA, Markhand GS, Mahar AR, Abro SA, Kanhar NA. Effect of water stress on yield and yield components of wheat (*Triticum aestivum* L.) varieties. Pak. J. Bot. 2009;41(3):1303-1310.

4. Poehlman J M, Sleper DA. Breeding Field Crops. 4th ed. Iowa State University Press, Ames, USA. 1995;494.

5. Khanna VK, Bajpai GC, Hussain SM. Effect of gamma radiation on germination and mature plant characters of wheat and triticale. Haryana Agricultural University Journal of Research. 1986;16(1):42-50.

6. Sobieh ESS. Induction of short culm mutants for bread wheat by using gamma rays. Arab Journal of Nuclear Sciences and Applications. 2002;35(1):309-317.

7. Al-Naggar AMM, Ragab AEI, Youssef SS, Al-Bakry RIM. New genetic variation in drought tolerance induced via irradiation and hybridization of Egyptian cultivars of bread wheat. Egypt. J. Plant Breed. 2004; 8:353-370.

8. Al-Naggar AMM, Atta MM, Shaheen AM, Al-Azab Kh F. Gamma rays and EMS induced drought tolerant mutants in bread wheat. Egypt. J. Plant Breed. 2007;11(3): 135-165.

9. FAO / IAEA. Mutant Variety Database. Cereals and Legumes. December, 2012. FAO/IAEA, Vienna. Available:http://mvgs.iaea.org.

10. Singh BD. Breeding for resistance to abiotic stresses. I. Drought resistance. In: Plant Breeding Principles and Methods. Kalayani Publishers, New Delhi, India. 2000;381-409.

11. Al-Naggar AMM, Abdel- Raouf MS, El-Borhamy HS, Shehab-El-Deen MT. Gene effects controlling inheritance of earliness and yield traits of bread wheat under drought stress conditions. Egypt. J. Plant Breed. 2012;16(3): 41- 59.

12. Al-Bakry MRI. Glaucous wheat mutants. I. Agronomic performance and epicuticular wax content. Egypt. J. Plant Breed. 2007;11(1):1-9.

13. Al-Naggar AMM, Shehab-El- Deen MT. Predicted and actual gain from selection for early maturing and high yielding wheat genotypes under water stress conditions. Egypt. J. Plant Breed. 2012;16(3):73 -92.

14. Tharwat E, Akbar H, Jaime A, Teixeira DS. Genetic analysis and selection for bread wheat (*Triticum aestivum* L.) yield and agronomic traits under drought conditions. International Journal of Plant Breeding. 2013;7(1):61- 68.

15. Xiao J, Li J, Yuan L, Mccouch, S, Tanksley SK. Genetic diversity and its relationship to hybrid performance and heterosis in rice as revealed by PCR-based markers. Theor. Appl. Genet. 1996; 92:637-664.

16. Ovesna J, Polakova K, Lisova L. DNA analysis and their applications in plant breeding Czech. J. Genet. Plant Breed. 2002;38: 29-40.

17. Momtaz OA, Hashem MM, Moghaieb REA, Hussein MH. Genetic polymorphism among Egyptian rice genotypes as revealed by RAPD, SSR and AFLP analyses. Arab J. Biotech. 2010;13(2): 173-184.

18. Akkaya MS, Shoemaker RC, Specht JE, Bhagwat AA, Cregan PB. Integration of simple sequence repeats DNA markers into a soybean linkage map. Crop Sci. 1995;35 :1439-1445.

19. Wang Z, Weber JL, Zhong G, Tanksley SD. Survey of plant short tandem DNA repeats. Theor. Appl. Genet. 1994;88:1-6.

20. Munir A, Armghan S, Iqbal M, Asif M, Hirani AH. Morphological and molecular genetic variation in wheat for salinity tolerance at germination and early seedling stage. AJCS. 2013;7(1):66-74.

21. Ivandiç V, Hackett CA, Nevo E, Keith R, Thomas WTB, Forster BP. Analysis of simple sequence repeats (SSRs) in wild barley from the fertile crescent: Associations with ecology, geography and flowering time. Plant Mol. Biol. 2002;48: 511–527.

22. Liviero L, Maestri M, Gulli E, Nevo N, Marmiroli E. Ecogeographic adaptation and genetic variation in wild barley, application of molecular markers targeted to environmentally regulated genes. Genet. Resources and Crop Evol. 2002;49:133–144.

23. Quarrie SA, Dodig D, Pekiç S, Kirby J, Kobiljski B. Prospects for marker-assisted selection of improved drought responses in wheat. Bulg. J. Plant Physiol. 2003;83-95.

24. Ciucǎ M, Petcu E. SSR markers associated with membrane stability in wheat (*Triticum aestivum* L.). Romanian Agricultural Research. 2009;26:21-24.

25. Abd El-Hadi AA. Molecular characterization of some durum wheat drought tolerant mutant by RAPD and ISSR analysis. Arab J. Biotech. 2012;15 (1):77-90.

26. El-Ameen T. Molecular markers for drought tolerance in bread wheat. African Journal of Biotechnology. 2013;12(21): 3148-3152.

27. El Siddig MA, Baenziger S, Dweikat I, El Hussein AA. Preliminary screening for water stress tolerance and genetic diversity in wheat (*Triticum aestivium* L.) cultivars from Sudan. Journal of Genetic Engineering and Biotechnology. 2013;2(2): 87-94.

28. Snedecor GW, Cochran WG. Statistical Method. 8th ed. Iowa State Univ. Press, Ames, USA; 1989.

29. Doyle JJ, Doyle JL. A rapid DNA isolation procedure for small quantities of fresh leaf. Phytochem. Bull. 1987;19:11-15.

30. Sumar A, Ahmet D, Gulay Y. Isolation of DNA for RAPD analysis from dry leaf materials of some *Hesperis* L. Speciments. Plant molecular Biology Reporter. 2003;21: 461-461.

31. Bousba R, Michael B, Abdelh AD, Samer L, Abdulkader D, Kadour B, Mustapha I, Gaboun F, Ykhlef N. Screening for drought tolerance using molecular markers and phenotypic diversity in durum wheat genotypes. World Applied Sciences Journal. 2012;16 (9):1219-1226.

32. Roider MS, Korzum V, Wendehake K, Plaschke J, Tixier M, Leroy P, Ganal MW. A microsatellite map of wheat. Genetics, 1998;149:2007-2023.

33. Sneath PHA, Sokal RR. Numerical Taxonomy. Freeman, San Francisco, California, USA. 1973; 513.

34. Sharma HP, Bhargava SC. Relative sensitivity of wheat genotypes under moisture stress conditions. Annals of Biology Ludhiana. 1996;12(1):39-42.

35. Ragab AI, Sobieh E-SSS. An attempt to improve bread wheat for water stress tolerance using gamma irradiation. Egypt. J. Appl. Sci. 2000;15(11):25-45.

36. Al-Bakry MRI, Al-Naggar AMM, Moustafa HAM. Improvement of grain yield of a glaucous wheat mutant line *via* backcrossing. Egypt. J. Plant Breed. 2008; 12(2):123-131.

37. Vega U, Frey KJ. Transgressive segregation in inter and intra-specific crosses of barley. Euphytica. 1980;29: 585-594.

38. Naghavi MR, Mohsen M, Ramshini HA, Bahman F. Comparative analysis of the genetic diversity among bread wheat genotypes based on RAPD and SSR markers. Iranian Journal of Biotechnology. 2004;2:195-202.

39. Hussein EHA, Abd- Alla SM, Awad Nahla A, Hussein MS. Assessment of genetic variability and genotyping of some Citrus accessions using molecular markers. Arab J. Biotech. 2003;7(1):23-36.

40. Moghaieb REA, Talaa NB, Abdel-Hadi AA, Youssef SS, El-Sharkawy AM. Genetic variation for salt tolerance in some bread and pasta wheat genotypes. Arab J. Biotech. 2010;13(1):125-142.

Permissions

List of Contributors

Md. Abul Hasanat
Department of Fisheries Biology and Genetics, Bangladesh Agricultural University (BAU), Mymensingh 2202, Bangladesh

Md. Fazlul Awal Mollah
Department of Fisheries Biology and Genetics, Bangladesh Agricultural University (BAU), Mymensingh 2202, Bangladesh

Md. Samsul Alam
Department of Fisheries Biology and Genetics, Bangladesh Agricultural University (BAU), Mymensingh 2202, Bangladesh

Chen Lanzhuang
Faculty of Environmental and Horticultural Science, Minami Kyushu University, 3764-1, Tatenocho, Miyakonojo City, Miyazaki, 885-0035, Japan

Nishimura Yoshiko
Faculty of Environmental and Horticultural Science, Minami Kyushu University, 3764-1, Tatenocho, Miyakonojo City, Miyazaki, 885-0035, Japan

Umeki Kazuma
Faculty of Environmental and Horticultural Science, Minami Kyushu University, 3764-1, Tatenocho, Miyakonojo City, Miyazaki, 885-0035, Japan

Zhang Jun
Qinghai Academy of Animal Science and Veterinary Medicine, Xining, Qinghai, 810016, China

Xu Chengti
Qinghai Academy of Animal Science and Veterinary Medicine, Xining, Qinghai, 810016, China

Komal Murtaza
Department of Genetics, Hazara University, Mansehra, Pakistan

Khushi Muhammad
Department of Genetics, Hazara University, Mansehra, Pakistan

Mukhtar Alam
Department of Plant Breeding and Genetics, University of Swabi, Swabi, Pakistan

Ayaz Khan
Department of Genetics, Hazara University, Mansehra, Pakistan

Zainul Wahab
Department of Art and Design, Hazara University, Mansehra, Pakistan

Muhammad Shahid Nadeem
Department of Genetics, Hazara University, Mansehra, Pakistan

Nazia Akbar
Department of Genetics, Hazara University, Mansehra, Pakistan

Waqar Ahmad
Department of Genetics, Hazara University, Mansehra, Pakistan

Habib Ahmad
Department of Genetics, Hazara University, Mansehra, Pakistan

M. Amdan
Laboratory of Virology, Microbiology and Quality / Ecotoxicology and Biodiversity, University Hassan II Mohammedia-FSTM, Morocco

H. Faquihi
Laboratory of Virology, Microbiology and Quality / Ecotoxicology and Biodiversity, University Hassan II Mohammedia-FSTM, Morocco

M. Terta
Laboratory of Virology, Microbiology and Quality / Ecotoxicology and Biodiversity, University Hassan II Mohammedia-FSTM, Morocco

M. M. Ennaji
Laboratory of Virology, Microbiology and Quality / Ecotoxicology and Biodiversity, University Hassan II Mohammedia-FSTM, Morocco

R. Ait Mhand
Laboratory of Virology, Microbiology and Quality / Ecotoxicology and Biodiversity, University Hassan II Mohammedia-FSTM, Morocco

Issam Saidi
Department of Biology, Laboratory of Plant Physiology and Biotechnology, Faculty of Sciences of Tunis, Campus Universitaire, 1060 Tunis, Tunisia

Nasreddine Yousfi
Department of Biology, Laboratory of Plant Extrêmophiles, Biotechnology Center of Borj Cédria, BP 901, 2050 Hammam-Lif, Tunisia

Wahbi Djebali
Department of Biology, Laboratory of Plant Physiology and Biotechnology, Faculty of Sciences of Tunis, Campus Universitaire, 1060 Tunis, Tunisia

Yacine Chtourou
Department of Biology, Laboratory of Toxicology and Environmental Microbiology, Faculty of Sciences of Sfax, Tunisia

Othman E. Othman
Department of Cell Biology, National Research Center, Dokki, Egypt

Mohamed F. Abdel-Samad
Department of Cell Biology, National Research Center, Dokki, Egypt

Nadia A. Abo El-Maaty
Department of Cell Biology, National Research Center, Dokki, Egypt

Karima M. Sewify
Department of Zoology, Girl Faculty, Ain Shams University, Egypt

K. K. Nkongolo
Department of Biology, Laurentian University, Sudbury, Ontario, P3E 2C6, Canada

G. Daniel
Department of Biology, Laurentian University, Sudbury, Ontario, P3E 2C6, Canada
Department of Chemistry and Biochemistry, Laurentian University, Sudbury, Ontario, P3E 2C6, Canada

K. Mbuya
National Maize Program, National Institute for Agronomic Study and Research, (INERA) B.P. 2037, Kinshasa 1, DR-Congo

P. Michael
Biomolecular Sciences Program, Laurentian University, Sudbury, Ontario, P3E 2C6, Canada

G. Theriault
Biomolecular Sciences Program, Laurentian University, Sudbury, Ontario, P3E 2C6, Canada

Alex Ngetich
Department of Biochemistry and Biotechnology, Kenyatta University, P.O.Box 43844-00100, Nairobi, Kenya

Steven Runo
Department of Biochemistry and Biotechnology, Kenyatta University, P.O.Box 43844-00100, Nairobi, Kenya

Omwoyo Ombori
Department of Plant Sciences, Kenyatta University, P.O.Box 43844-00100, Nairobi, Kenya

Michael Ngugi
Department of Agriculture, Meru University College of Sciences and Technology, P.O.Box 972- 60200, Meru, Kenya

Fanuel Kawaka
Department of Pure and Applied Sciences, Technical University of Mombasa, P.O.Box 90420, 80100, Mombasa, Kenya

Arusei Perpetua
Department of Botany, Moi University, P.O.Box 3900, 30100, Edoret, Kenya

Gitonga Nkanata
Department of Agriculture, Meru University College of Sciences and Technology, P.O.Box 972- 60200, Meru, Kenya

Lawan Abdu Sani
Department of Plant Science, Ahmadu Bello University, Zaria, Nigeria
Department of Plant Biology, Bayero University Kano, Nigeria

Inuwa Shehu Usman
Department of Plant Science, Ahmadu Bello University, Zaria, Nigeria

Muhammad Ishiaku Faguji
Department of Plant Science, Ahmadu Bello University, Zaria, Nigeria

Sunusi Muhammad Bugaje
Department of Plant Science, Ahmadu Bello University, Zaria, Nigeria

Olufemi O. Oyelakin
Biotechnology Centre, Federal University of Agriculture, Abeokuta, Nigeria
Central Biotechnology Laboratory, International Institute of Tropical Agriculture (IITA) Ibadan, Nigeria

Jelili T. Opabode
Department of Crop Production and Protection, Obafemi Awolowo University, Ile-Ife, Nigeria

Emmanuel O. Idehen
Department of Plant Breeding and Seed Technology, Federal University of Agriculture, Abeokuta, Nigeria
Biotechnology Centre, Federal University of Agriculture, Abeokuta, Nigeria

Ishaya Y. Longdet
Department of Biochemistry, University of Jos, Nigeria

Hajiya M. Inuwa
Department of Biochemistry, Ahmadu Bello University, Zaria, Nigeria

Isma'il A. Umar
Department of Biochemistry, Ahmadu Bello University, Zaria, Nigeria

Andrew J. Nok
Department of Biochemistry, Ahmadu Bello University, Zaria, Nigeria

Olalekan Ibrahim Sobowale
Crop Improvement Division, Cocoa Research Institute of Nigeria, Ibadan, Nigeria

Benjamin Oluwole Akinyele
Department of Crop, Soil and Pest Management, The Federal University of Technology Akure, Ondo State, Nigeria

Daniel Babasola Adewale
Department of Biological Sciences, Ondo State University of Science and Technology, Okitipipa, Ondo State, Nigeria

Balwinder Singh
Plant Tissue Culture Laboratory, Department of Biotechnology, Khalsa College Amritsar, 143 002 (Punjab), India

A. M. M. Al-Naggar
Department of Agronomy, Faculty of Agriculture, Cairo University, Giza, Egypt

Kh. F. Al-Azab
Department of Plant Research, Nuclear Research Center, Atomic Energy Authority, Inshas, Egypt

S. E. S. Sobieh
Department of Plant Research, Nuclear Research Center, Atomic Energy Authority, Inshas, Egypt

M. M. M. Atta
Department of Agronomy, Faculty of Agriculture, Cairo University, Giza, Egypt

Printed in the USA
CPSIA information can be obtained
at www.ICGtesting.com
JSHW051447221024
72173JS00006B/1600

9 781682 860069